T0275987

Bioengineering and Cancer Stem Cell Concept

Mirjana Pavlovic • Bela Balint

Bioengineering and Cancer Stem Cell Concept

 Springer

Mirjana Pavlovic
Department of Computer and Electrical
Engineering and Computer Science
Florida Atlantic University
Boca Raton, FL, USA

Bela Balint
Military Medical Academy
Institute for Hematology and Transfusiology
Belgrade, Serbia

ISBN 978-3-319-25668-9   ISBN 978-3-319-25670-2   (eBook)
DOI 10.1007/978-3-319-25670-2

Library of Congress Control Number: 2015957195

Springer Cham Heidelberg New York Dordrecht London

Printed on acid-free paper

Springer International Publishing AG Switzerland is part of Springer Science+Business Media
(www.springer.com)

This book is dedicated to the shadows of Dr. Joseph Caruso, friend, passionate scientist, and great human being, and Kristina and Janosh Balint, the wonderful parents who left so much after they perished.

Thank You Note

This book is the product of interaction between these author's mind and those who were always around in order to help and interact with penetrating discussions, supportive attitude, and friendly optimism indicating that there is a light at the end of this road less traveled and that it will enlighten not only those who decided to write but also those who will share the content of that writing once it is published.

From my side to my colleagues, friends, my wonderful students, and all of those who never forgot that giving is the highest expression of power and love! To Dr. Maria Larrondo Petrie, Dr. Zvi Roth, Dr. Borko Furht, Dr. Alex Kotlarchyk, Dr. Sharmistha Chatterjee, Sandhya Sharma, Jennifer Tarakmi, Small Susaye, Shilla Knutsen, Eleonora Trnovski, Zurima Luff, John Myfield, Karl Schreiber, Chad Coarsey, Charles Weinthal, and all of those who followed the evolution of this book with their active input, love, and affection.

Mirjana Pavlovic

Preface

The concept of cancer stem cell (CSC) is a topic and very attractive field in cancer biology due to its possible broad impact onto cancer treatment options and diagnostic procedures. Although the theoretical explanations of cancer stem cell involvement in leukemia and solid cancers are still full of controversies with two leading theories (hierarchical and stochastic/cancer stem cell model) being investigated and checked, there is less and less doubts that such an entity as cancer stem cell does exist within otherwise heterogeneous cancer cell population.

Inspired 150 years ago by the famous German scientist Rudolph Virchow (1863), the development of novel approaches involving bioengineering solutions into both the diagnostics and the treatment is of critical importance today. The methods for isolation, determination of critical markers for particular cancers, and confirmation of functionality/tumorigenicity are developed in the meanwhile. The features of cancer stem cells are now compared to those of normal stem cells and the differences and similarities, investigated. The book will deal with novel methodology for: cancer stem cell marker identification, purification of CSCs, and operation of critical metabolic pathways in cancer stem cells, their genetic and epigenetic mechanisms emphasizing DNA methylation. The concept of targeted cancer stem cell therapy will be revealed from diagnostic, prognostic, and curative aspects. The ultimate goal of the book is to help the starters in this field to find the most critical information at one place and condensed, reliable source of references for intended research, as well as to get informed with the newest concepts and approaches linked to this topic.

Boca Raton, FL   Mirjana Pavlovic

Contents

1 Normal Stem Cell: Entity or State? 1

Organogenesis from Adult Stem Cells and Problems
with Different Tissues 4

Therapeutic Implications for TCSCs as a New Concept 5

The Concept of VSEL 7

The Concept of Mesenchymal Stem Cell (with Dental Pulp
Cells as an Example) 9

Mobilization as a New Noninvasive Therapeutic Concept 12

Emphasizing Bioengineering Aspects to Stem Cell Engineering 13

Directions and Relevant Studies: We and Others 13

    What Is "Stemness?" 13

Breakthrough: Induced Pluripotent Stem Cells 16

Reprogramming as a Therapeutic Event 17

References 19

**2 Normal Stem Cells: Biology, Collection/Harvesting,
and Ex Vivo Manipulations** 25

Introduction 25

Bone Marrow Derived Stem Cell Collection 28

    Peripheral Blood Derived Stem Cell Harvesting/Purification 30

Umbilical Cord Blood Derived Stem Cell Usage 35

Long-Term Storage of Stem Cells by Cryopreservation 36

References 39

3 A Concept of Cancer Stem Cells: Entity and Theories 43

Introduction 43

What Are Cancer Stem Cells? 43

Isolation and Characterization of Stem Cells 46

Novel Treatment Modalities 50

Conclusions 53

References 54

**4   Cancer Stem Cell Markers: Classification
     and Their Significance in Cancer Stem Cells**   57
     Introduction   57
     Background and Significance   58
     Cancer Stem Cell Markers: Classification   60
     Are Cancer Cells Transformed Normal Stem Cells?   60
     Cancer Stem Cell Markers and Their Function in Normal
     and Cancer Stem Cells   62
     Conclusions   62
     References   68

5   Genetic Mechanisms Involved in Cancer Stem Cell Emergence   71
     References   72

6   Epigenetic Mechanisms Involved in Cancer Stem Cell Profiles   73
     Introduction   73
     Background and Significance   75
     Genetic and Epigenetic Phenomenon in Cancer Development   76
     Cancer Stem Cell Concept   77
     Cancer Stem Cell Models: Theories and Possible Explanations   78
     Cancer Stem Cell Markers: Conceptual Features   79
        Significance of Computational Approaches for Cancer
        Prediction: Identifying Methylation Profiles in Cancer Stem Cells   80
     Conclusions   84
     References   84

7   Mitochondrial Respiration of Cancer Stem Cell   89
     Introduction   89
     The Role of Mitochondria in Energy Metabolism:
     Oxidative Phosphorylation and Uncoupling Effect, Apoptosis,
     Calcium, and ROS Balancing   89
     The Change in Lipid Content on the Inner Mitochondrial
     Membrane and Uncoupling Effect of Mitochondria
     in the Cancer Cells   91
     Therapeutic Approaches: Targeting Mitochondria
     in Cancer Stem Cells   91
     Conclusions   92
     References   94

8   Metabolism in Cancer Stem Cell   97
     References 101

9   Different Approaches for Anticancer/Antitumor Therapy 103
     Introduction 103
     Cancer-Targeted Therapy 103
        Antibody-Targeted Therapy 104
        Small Molecules-Targeted Therapy 104

Ligand-Targeted Therapy... 106
Kinase Inhibitors Targeting Multiple Signaling Pathways 107
Attacking Multiple Targets 108
Agents Targeting Serine/Threonine Kinases....................................... 109
Heat Shock Protein Targets.. 110
Cancer Stem Cells Therapy: Concept and Novelties 111
Possibilities of Engineering Targeted Cancer SC Therapy
Using Principles of Magnetism.. 112
Cancer Therapy Outlook.. 116
    Chemoresistance, Radioresistance, and Reactive Oxygen
    Species as Targets 116
Conclusions.. 118
References.. 118

10 **Targeted Cancer Stem Cell Therapy**... 123
General: Remote Control of Targeted Drug Delivery
as Possibility for Cancer Stem Cell Therapy 123
    Nanotechnology/Nanoparticles.. 124
    Biomagnetism/Magnetic Particles 126
    Immunotherapy (Rational Vaccine Design) for CSCs 126
    Epigenetic Synthetic Lethal Therapeutic Approaches
    to ARID1A Mutated Cancers.. 126
    Genetic Engineering-Based Therapy: Engineered Normal
    (Toxic) Stem Cells as a Therapeutic Tool for Cancer.......................... 127
Conclusions.. 128
References.. 128

11 **Bioengineered CSC Tumors**... 133
Introduction... 133
First Tumor Models Based on CSCs.. 134
3-D Printing as "Universal" Bioengineering Method 134
3-D Printed Tumor Models: State of the Art 135
References.. 137

12 **Summary on the Role of Bioengineering in Cancer
    Stem Cell Paradigm** 139
Stem Cell Paradigm: Resume 139
Self-Renewal of Stem Cells 141
The Role of Bioengineering in Understanding
CSCs Paradigm: Resume 142
The Concept of Targeted CSC Therapy: Resume.................................... 142
References.. 143

Chapter 1
Normal Stem Cell: Entity or State?

That which can be asserted without evidence, can be dismissed without evidence.

Christopher Hitchens

The first use of stem cells in humans was done by physicians who were tempted to use them in trying to treat hematological disorders. Stem cell transplantation was pioneered using bone marrow-derived stem cells by a team at the Fred Hutchinson Cancer Research Center from the 1950s through the 1970s led by Edward Donnall Thomas, whose work was later recognized with a Nobel Prize in Physiology or Medicine [1]. Thomas' work showed that bone marrow cells infused intravenously could repopulate the bone marrow and produce new blood cells. His work also reduced the likelihood of developing a life-threatening complication called Graft versus Host Disease (GvHD). The first physician to perform a successful human bone marrow transplant was Robert A. Good at the University of Minnesota in 1968 [1]. With the availability of the stem cell growth factors (GM-CSF and G-CSF), most hematopoietic stem cell transplantation procedures are now performed using stem cells collected from the peripheral blood, rather than from the bone marrow. Collecting peripheral blood stem cells provides a bigger graft, does not require the donor to be subjected to general anesthesia in order to collect the graft, results in a shorter time to engraftment, and may provide for a lower long-term relapse rate.

The first recorded attempt at cellular therapy occurred in 1912 when German physicians attempted to treat hypothyroid children with thyroid cells. Cellular therapy, as practiced today, was developed in the early 1930s by Paul Niehans M.D. (1882–1971), a Swiss physician who became known as "the father of cell therapy." It soon became popular with celebrities as a means of rejuvenation. A 1990 article in In Health magazine described Niehans as a "public relations genius" and stated that the Clinique La Prairie, which he had founded in Clarens-Montreux, Switzerland, had attracted 65,000 patients. Its 1999 1-week "revitalization program" costed about $8000 [1].

Generally, the Stem Cell (SC)—compartment is divided into embryonic and tissue-specific or adult SCs [1]. Embryonic SCs (ES or ESC) are by definition the "master cells" with the largest spectrum of differentiation potential, e.g., capable of differentiating into every type of cells either in vitro or in vivo. Thanks to the presence of embryonic body, these cells have ability to develop into three primary layers: endo-

© The Author(s) 2015
M. Pavlovic, B. Balint, *Bioengineering and Cancer Stem Cell Concept*,
DOI 10.1007/978-3-319-25670-2_1

1

derm, ectoderm, and mesoderm [1]. The discovery of SCs inside cell mass of embryos and in adult tissue has revolutionized the medical field by introducing new therapeutic dimensions into previously untreatable diseases and injuries. Several experimental or preclinical studies have suggested that application of ESC could be promising in the treatment of various diseases [2–6]. However, recognition of appropriate ethical aspects, regulatory acts, and standardization in ESC-mediated regenerative medicine is needed as it is still the matter of controversy. Besides, permanent, persistent, and accurate updating of the facts regarding their phenotypic, functional, and immunologic characteristics is an essential requirement for safe clinical application of SCs. Some authors stand that the initial theory that ESCs are ignored by immunocompetent hosts was overlooked. On the contrary, they think that it is even more evident that ESCs could protect themselves actively by several immunomodulatory mechanisms against T lymphocytes and natural killer cells of host, and actively participate in immune-mediated events. Recent isolation of fetal SCs from several sources either at the early stages of development or during the later trimesters of gestation, sharing similar growth kinetics and expressing markers of pluripotency, provides strong support to the statement that these cells may be biologically closer to ESCs. In fact, they represent intermediates between embryonic and adult mesenchymal SCs with regards to proliferation rates and plasticity features, thus being able to confer an advantage over postnatal mesenchymal SCs derived from conventional adult sources.

Bone marrow was the primary source of SCs for transplant [1]. However, peripheral blood and umbilical (cord) blood are also currently used as sources. SCs derived from these sources may have therapeutic potential (without severe adverse effects) only when given to the individual from whom they were derived (autologous transplants) or from an immunologically matched donor (allogeneic transplants) [1].

Despite the fact that the ideal type and source of cells have not yet been defined, immature SCs are capable of colonizing different tissues due to ability of homing and transdifferentiation or lineage plasticity, in the settings of regenerative medicine. Furthermore, there are several facts suggesting that adult SCs and even differentiated somatic cells, under appropriate microenvironmental cues or signals, are able to be "reprogrammed" and contribute to a much wider spectrum of differentiated progeny than previously anticipated. This has been demonstrated by using tissue-specific SCs—which like ESCs—do not express CD45 as an exclusive hematopoietic marker. Consequently, adult mesenchymal SCs and endothelial precursors seem to be clinically applicable for cell-mediated, regenerative therapy of patients with myocardial, brain, vascular, liver, pancreas, and some other tissue damages.

It is widely accepted that allogeneic transplants are still the most efficient treatment for patients with liver failure and Chronic Myelogenes Leukemia (CML) [1]. However, there is a lack of donors and some alternative therapeutic approaches are therefore needed. Transplantation of mature hepatocytes has been evaluated, but the long-term efficacy remains unclear and the paucity of donor cells makes this strategy quite limited. The use of SC-therapy transplantation is perhaps a more promising alternative approach.

The intensification of myeloablative radio-chemotherapy enlarged the use of SC transplants, as well as the introduction of cell-mediated therapeutic approaches in regenerative medicine resulting in increased needs for both specific blood-derived progenitor/cells, and practical operating procedures inducing minimized cellular damages during their collection or processing and storage in frozen state. Therefore, successful performance of SC transplants or cell-mediated therapy requires efficient collection, processing, and (cryo) preservation procedures for obtaining an acceptable cell yield and post-thawing recovery, as well as advantageous clinical outcome. For wound healing in the skin, epidermal stem cells and bone marrow progenitor cells both contribute. Thus, it is likely that organ-specific progenitors and hematopoietic stem cells are involved in repair, even for other organ repair. In summary, stem cells could be described as:

- Foundation cells for every organ, tissue, and cell in the body.
- A "blank microchip" that can ultimately be programmed to perform any number of specialized tasks.
- Undifferentiated "blank" cells that do not yet have a specific function.
- Self-sustaining and capable of replicating themselves for long periods of time.
- Under proper conditions, begin to develop into specialized tissues and organs [1].

These unique characteristics make stem cells very promising potential for supplying cells and tissues instead of organs in a spectrum of devastating diseases from diabetes type 1 to stroke, spinal cord injuries, and myocardial infarction [1–7]. In the situation when the number of people needing organ and tissue transplants exceed the number of donated organs and tissues, this is the promise and hope, which deserves a deep and serious consideration. However, despite rapidly growing knowledge on adult stem cell sources, features, and use, there are still some fundamental remaining questions regarding them that include: Does only one common type of stem cell migrate to different organs and repair tissue or are there multiple types of stem cells? Does every organ have stem cells (some of which have not yet been discovered)? Are the stem cells programmed to divide a finite number of times or do they have unlimited cell proliferation capacity? According to their functionality, stem cells can be divided in two categories: normal and cancer stem cells [1].

1. *Normal stem cells (embryonic, cord blood, or adult)* are immature cells that can replicate, or renew them, and are able to differentiate, or mature into all the cells that an organism or particular organ system needs. In other words, they possess a kind of immortality marked as self-renewal because these cells can divide indefinitely to produce more copies of them. Each stem cell is unspecialized, but it can produce progeny that mature into the various cell types of, say, the brain or the immune system. Once this maturation occurs, these adult stem cell heirs may divide rapidly but only a limited number of times [1–7]. The primary purpose of adult stem cells is healing. Finding out how adult stem cells store information and transform themselves into other cells with different properties is a fascinating topic for exploration [8–13]. Stem cells are so named because cells are derived from a main stem or mother set of cells. This is similar to a tree trunk that provides the stem from which other cells grow and branch out into other types of cells.

2. *Cancer stem cells.* Finding cancers' stem cells is a rapidly growing area of research [5, 7, 14–17]. These cancer-causing cells, which make up a tiny

fraction of cells within tumors, have properties similar to those of stem cells [5]. Cancer stem cells make up only a tiny number of the total cancer cells in a leukemia patient, which makes the cells next to impossible to find. Therefore, it seems that promise of this line of research can only be realized, by studying adult stem cells as well as embryonic stem cells (ES). The latter are still ethical problem and therefore substantially controversial because an early embryo is destroyed when researchers remove stem cells from it.

An alternative is to take the stem cells from embryos that carry a genetic defect for specific diseases. Are cancer cells transformed normal stem cells? Researchers have traditionally thought of cancer as a collection of cells, all growing exponentially. According to the new research, conventional cancer therapies do an effective job killing the majority of cells within the tumor, but they may miss cancer stem cells. As a result, cancers often reoccur. Even hematologic and some non-hematologic malignancies treated by autologous stem cell transplant and high dose chemotherapy have shown that regardless of survival rate of some cancers, the final outcome is death, due to recurrence of cancer. The reason is (among others) in the fact that clinicians are injecting also cancer cells with healthy stem cells during reinfusion after apheresis collection, which accumulate and renew with a time to the critical level causing relapse or death. Ontogeny (development of an organism) and oncology (cancer development) share many common features. From the 1870s the connection between development and cancer has been reported for various types of cancers [1]. Existence of "cancer stem cells" with aberrant cell division has also been reported more recently [5]. The connection between cancer and development is clearly evident in teratocarcinomas. As early as 1862, Virchow discovered that the germ cell tumor teratocarcinoma is made up of embryonic cells [1]. In 1970, Stevens derived embryonic carcinoma cells from teratocarcinoma (a spontaneous tumor of germ cells that resembles development gone awry) [1]. This tumor may contain several types of epithelia: areas of bone, cartilage, muscle, fat, hair, yolk sac, and placenta. These specialized tissues are often adjacent to an area of rapidly dividing unspecialized cells. The teratocarcinomas are able to differentiate into normal mature cells when transplanted into another animal. This alternation between developmental and tumor cells status demonstrates how closely development and cancer are related. The present-day challenge is to decode the common molecular mechanism and genes involved in self-renewal for cancer cells and stem cells.

Organogenesis from Adult Stem Cells and Problems with Different Tissues

How do a small number of stem cells give rise to a complex three-dimensional tissue with different types of mature cells in different locations? This is the most fundamental question in organogenesis. The hematopoietic and nervous systems employ very different strategies for generating diversity from stem cells. The hematopoietic system assiduously avoids regional specialization by stem cells.

Hematopoietic stem cells are distributed in different hematopoietic compartments throughout the body during fetal and adult life, and yet these spatially distinct stem cells do not exhibit intrinsic differences in the types of cells they generate. This contrasts with the nervous system, where even small differences in position are associated with the acquisition of different fates by stem cells.

While local environmental differences play an important role in this generation of "neural diversity," we must accept that intrinsic differences between stem cells are also critical. Part of the reason why different types of cells are generated in different regions of the nervous system is that intrinsically different types of stem cells are present in different regions of the nervous system. To understand the molecular basis for the regional patterning of neural stem cell function, we are now studying how these differences are encoded.

Therapeutic Implications for TCSCs as a New Concept

To prove the stem cells derived from bone marrow (BM) and peripheral blood, including hematopoietic stem cells, are indeed transformed into solid-organ-specific cells, several conditions must be met:

1. The origin of the exogenous cell integrated into solid-organ time must be documented by cell marking, preferably at the single-cell level.
2. Cell should be processed with a minimum of "ex vivo" manipulation (e.g., culturing) which may make them more susceptible to crossing lineages.
3. The exogenous cells must be shown to have become an integral morphologic part of the newly acquired tissue.
4. Transformed cells must be shown to have acquired the function of the particular organ into which it has been integrated both by expressing organ-specific proteins and by showing specific organ function.

Organ/tissue-specific niche (like in BM and liver)—exists as a deposit (storage) of the adult stem cells in a specific location (Fig. 1.1). These cells are circulating in a very low number in the blood [18]. Accumulating evidence suggests that stem cells may also actively migrate/circulate in the postnatal period of life. Stem cell trafficking/circulation may be one of the crucial mechanisms that maintains the pool of stem cells dispersed in stem cell niches of the same tissue, that are spread throughout different anatomical areas of the body. This phenomenon is very well described for HSC, but other, already tissue committed stem cells (TCSC) (for example, endothelial, skeletal muscle, skeletal or neural stem cells) are probably circulating as well [18].

BM is the home of migrating stem cells with not only hematopoietic stem cells within their niches, but also a small number of TCSC, which might be the reason why many authors think that HSC may transdifferentiate, although we do not have a direct proof for that. They might have plasticity, but not necessarily the "transdifferential" potential [18]. What is differentiated in the tissue of injection might be TCSC characteristic for that tissue. It has been shown that number of these cells is decreased with

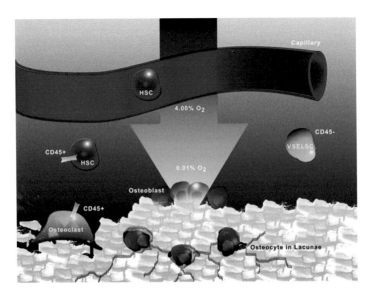

Fig. 1.1 Differences in phenotypes (external and internal markers) between HSC and VSEL from mouse bone marrow

ageing (long living and short living mice and humans). It would be interesting to identify genes that are responsible for tissue distribution/expansion of TCSC. These genes could be involved in controlling the life span of the mammals. Therefore, BM stem cells are a heterogeneous population of cells with HSC and TCSC, the morphological and functional characteristics of which are different from HSC. Their number among BM MNC is very low (1 cell per 1000–10,000 BM MNC) within young mammals and might play a role in small injuries [1]. In severe injuries like heart infarct or stroke they have no possibility to reveal their full therapeutic potential. The allocation of these cells to the damaged areas depends on homing signals that maybe inefficient in the presence of some other cytokines or proteolytic enzymes that are released from damaged tissue-associated leukocytes and macrophages [13]. We can envision, for example, that metalloproteinases released from inflammatory cells may degrade SDF-1 locally, and thus perturb homing of CXCR4+TCSC. There is possibility that these cells while "trapped" in BM are still in: "dormant" stage-not fully functional, and need the appropriate activation signals by unknown factors [18]. These cells also, at least in some cases could be attracted to the inflammatory areas, and if not properly incorporated into the damaged tissue they may transform and initiate tumor growth. In summary, between the pools of TCSC, there are probably those already committed to transdifferentiate into neural cells, or cells of tissues and organs other than neural, but we still do not have the control over their tracking, homing, and finally regenerative capacity in the given tissue, which is a fundamental prerequisite for successful regenerative therapy [8–13].

The Concept of VSEL

In a discovery that has the potential to change the face of stem cell research, a University of Louisville scientist has identified cells in the adult body that seem to behave like embryonic stem cells [18–45]. The cells, drawn from adult bone marrow, look like embryonic stem cells and appear to mimic their ability to multiply and develop into other kinds of cells. The finding, presented the first time at the 47th Annual Meeting of the American Society of Hematology (ASH) in Atlanta, was announced on December 12 at the society's news conference. A study by Ratajczak's team published in 2005 year in the journal "Leukemia" was the first to identify a type of stem cell in adult bone marrow that acts differently than other marrow stem cells [18]. The newly identified cells, called "very small embryonic-like" (VSEL) stem cells, have the same ultrastructure and protein markers as embryonic stem cells [18–40]. Ratajczak and several other researchers from University of Louisville in the presentation at the ASH meeting showed that VSEL stem cells mobilize into the bloodstream to help repair damaged tissue following a stroke [30]. In further research advance, Ratajczak's team also has grown VSEL cells in a lab and has stimulated them to change into nerve, heart, and pancreas cells [30]. The difference in markers between HSC and VSELs in mouse is shown in (Fig. 1.1), while the differences in ultrastructure are shown in (Fig. 1.2).

Along with this new concept, there is a premise that in regenerative therapy done before, with hematopoietic stem cells (considered to have plasticity and multipotency) the VSELs were "contaminants" that actually contributed to positive regenerative clinical outcome, since they have those capabilities [18]. This is an interesting concept which should be seriously considered in humans.

Thus, since VSELs have been found in human cord blood and bone marrow, they seem to be of a critical importance for consideration of stem cell transplant choice based upon the phenotype and number of stem cells aimed to be transplanted within

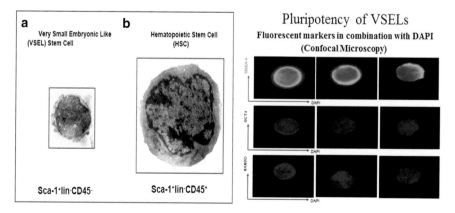

Fig. 1.2 Differences in ultrastructure between HSC and VSEL. (**a**) Very Small Embryonic-Like (VSEL) Stem Cell. (**b**) Hematopoietic Stem Cell

a given clinical scenario. Despite conflicting data about this population [43–45], they are getting more confirmation in scientific community [31–40]. These cells have a great potential and like induced stem cells, can potentially eliminate the need for embryonic stem cells given that in adult organism they have all necessary components (parameters) that embryonic cells have, with a highest potency for lineage differentiation [41, 42].

1. Morphological studies have discovered that VSELs are unusually small (3–4 μm) eukaryotic cells which do possess several features of embryonic cells. Thus, the strategy based on FACS sorting of these cells should consider whether other adult tissues have those primitive little cells bigger than thrombocytes but smaller than erythrocytes [40–46].

2. These cells also express high nucleo/cytoplazmatic ratio and smaller cytoplazmatic region compared to HSCs and mature granulocytes. Beside the fact that it has confirmed the features such as size, confocal microscopy has also confirmed that VSELs express Oct-4, a hallmark of pluripotency of embryonic stem cells. In sum, morphological studies have discovered that VSELs are unusually small eukaryotic cells with several fundamental features of embryonic stem cells except tumorigenicity (pluripotency, sphere formation, embryonic bodies, and small size) [28].

3. These cells in a suitable medium perpetuate self-renewal longer, without jumping into differentiation, while on the other side they are capable of differentiating into bigger number of cell types in a suitable/conditioned medium into most of the cell (pancreatic cells, neural cells, cells of heart muscle and liver) which makes them suitable for expansion and reparative and regenerative purposes [28–30].

4. VSEL cells are, accordingly, a unique and distinguished entity rather than state with the features of plasticity, that questions plasticity of HSCs, suggesting strongly that that particular feature of BM stem cells could be in essence artifact caused by contamination of VSELs. Finally, the discovery of VSELs in the CB, PB, and BM of humans indicate their significance with respect to other features. Some other researchers before Ratajczak have not succeeded to completely isolate this fraction [46–48], probably due to bad technique of isolation and timing. More extensive and deeper studies in the future will show what is true and possible.

5. Key advantages associated with VSELs seem to be that they avoid the ethical or moral dilemmas associated with the use of embryonic and fetal cells, the potential negative biological effects associated with ESCLs such as their propensity for tumor formation, and the use of autologous stem cells to avoid immune rejection.

6. The studies on mouse model suggest necessity for the human studies on VSELs since it would be of great interest to check if these intriguing population of stem cells are also involved in caloric intake, longevity, and regenerative features of this distinctive stem cell entity [19]. While this book was prepared for print a recent report from Ratajczak's group in the form of Editorial explained

Table 1.1 VSELs: pros and cons with respect to different findings

	Authors:	Authors:
Parameters of VSELs necessary to detect in order to be able to consider their pluripotent function	Cons Dulak, May 2013, Weissman, Augt, 2013	Pros Kassmer et al. Dec, 2013, Bhartya et al. 2011–2013, Wang J, X. Guo et al. 2013, W. Wojakowski, 2013, R. Taichman, 2013, Chang et al. 2013, Havens et al. 2013
DNA amount	Little	Abundant
Formation of spheres	No	Yes
Octapeptide-4 expression	No	Yes
Differentiation into other lineages/blood cells	No	Into epithelial cells and cardiac cells, multipotent tissue progenitors in vitro and in vivo

Note: The other two authors who commented existence of VSELs in negative sense were Alison Abbot (Nature, 2013) who gave short reviews on the matter but not her own results, and Paul Koepfler who initiated negative comments on his blog (www.ipscell.com/tag/russell-taichman/)

many aspects of conflicting data in VSELs history in a very professional way strongly suggesting that VSELs are rather detectible entity than the state of stem cell (Table 1.1) [46–50].

The Concept of Mesenchymal Stem Cell (with Dental Pulp Cells as an Example)

Many human tissues are the source of stem cells responsible for tissue development and regeneration. Beside BM (Bone Marrow Stromal Stem Cells—BMSCs), currently it is considered that dental pulp is practically the most approachable and the most important source of adult mesenchymal stem cells [49–54]. Within the last 8 years, several populations of stem cells from dental pulp were isolated and characterized: (1) Dental Pulp Stem Cells—DPSCs), (2) Stem Cells From Human Exfoliated Decidual Teeth (SHEDs), and (3) Immature Dental Pulp Cells (IDPCs) [51–54]. These cells are of the ectomesenchymal origin, located in perivascular niche, highly proliferative, clonogenic, multipotent, and similar to BMSCs.

In in vitro conditions, they can differentiate with certain intercellular differences toward odontoblasts, chondrocytes, osteoblasts, adipocytes, neurons/glial cells, and smooth and skeletal muscle cells. In in vivo conditions, after implantation, they show different potential for dentine formation, as well as osteogenesis; after transplantation in mouse with compromised immune system, they make good grafts in different tissues and are capable of migrating into the brain, where they survive a certain time while reaching neurogenic phenotype. DPSCs have immunomodulatory effect, as they can be involved into immune response during infection of dental pulp by NF-kB activation, and by inhibiting T-lymphocyte proliferation, suggesting their immunosuppressive effect [51–54]. The future research should give us the complex data on the molecular and functional characteristics of dental pulp stem cells, as well as differences between different populations of these cells. Such research would fundamentally contribute to the better knowledge on the dental pulp stem cells, which is necessary due to their potential clinical application in in vivo cell transplantation, tissue engineering, and gene therapy (in vivo and ex vivo). Actually, by the isolation of IDPCs, which are the most primitive, but also the most plastic, (similar to embryonic stem cells), they are opening the new perspectives in a potential therapeutic application of these cells not only in regeneration of dentine, but also the regeneration of periodontal tissue and bone-junctional tissue of cranio-facial region, as well as in the therapy of neurotrauma, myocardial infarction, and connective tissue damage (Table 1.2, Fig. 1.3).

However, the shift of the logic and turning of the sense, entitling phenotypically defined populations as stem cells (although only some of them within that "cluster" are stem cells indeed), have introduced so much confusion into this discipline, that it is very difficult to perform corrections nowadays. It is at the same time the reason why many discoveries that enable stem cell therapy on the rodents, do not work on humans. One has to be very critical with respect to stem cell markers and its functional properties in order not to make a mistake in stem cell therapy (Fig. 1.4).

Table 1.2 Expression of protein or gene profiles in some dental stem cells in in vitro cultivation and relationship toward BMSC

Antigen	DPSC[a]	SHED	PDLSC	BMSC
CD14	–	–	–	–
CD34	–	–	–	–
CD44	++	++	++	++
CD45	–	–	–	–
CD106	+	+/–	+/–	++
CD146	++/+/–	++/+/–	++/+/–	++/+/–
3G5	+/–	+/–	+/–	+/–
Stro-1	++/+/–	++/+/–	++/+/–	++/+/–
α-Smooth muscle actin	++/–	++/–	++/–	++/+/–
Collagen type-I	++	++	++	++
Collagen type-III	++/+	++/+/–	++/+/–	++/+
Alkaline phosphatase	++/+/–	++/+/–	++/+/	++/+/–
Osteocalcin	++/+	++/+/–	++/–	+/–
Osteonectin	++/+	++/+	++/+	++/+
Osteopontin	+/–	+/–	+/–	+/–
Sialoprotein of the bone	–	–	–	–
Skleraksis	+	+	++	+
Sialophosphoprotein of the dentine	–	–	–	–

[a]*DPSC* Dental Pulp Stem Cells, *SHED* Stem Cells From Human Exfoliated Decidual Teeth, *PDLSC* Periodontal Ligament Stem Cells, (++) strong expression, (+) weak expression, (–) negative, (/) subpopulation

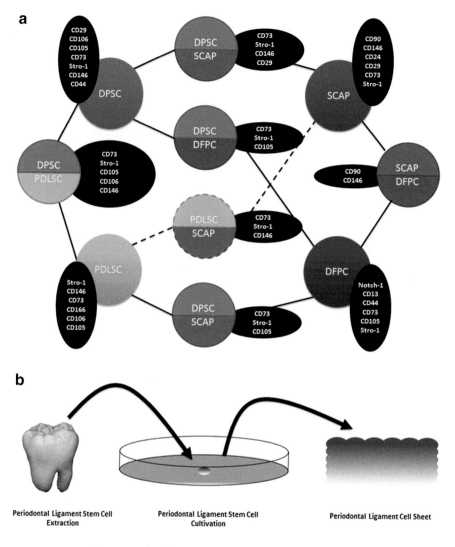

DPSC-Dental Pulp Stem Cell; DFPC-Dental Follicular Precursor
Cell; SCAP-Stem Cell of Apical Papilla; PDLSC-Periodontal
Ligament Stem Cell

Fig. 1.3 (**a**) and (**b**). The most important superficial cellular markers of dental pulp stem cell according to *DPSC* Dental Pulp Stem Cell, *DFPC* Dental Follicular Precursor Cell, *SCAP* Stem Cell of Apical Papilla, *PDLSC* Periodontal Ligament Stem Cell, according to Morsczeck et al. Clin Oral Invest 2008; 12:113–8

MSC (CD34⁻) IMI Belgrade   MSC (K67⁺)

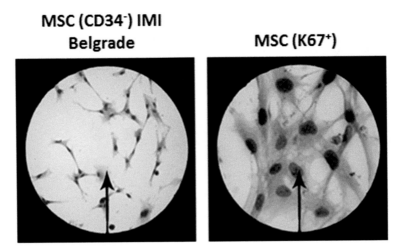

Fig. 1.4 Mesenchymal stem cells (With courtesy of Prof. Dr. Vera Todorovic, Institute for Medical Research, Belgrade, Serbia)

Mobilization as a New Noninvasive Therapeutic Concept

The classification of patients into "good" or "poor" mobilizers is based on CD34+ cell count in their peripheral blood (PB) after granulocyte-colony-stimulating factor (G-CSF) injection. CD34+ cells mobilized into peripheral blood (PB) are considered a more convenient source of hematopoietic stem and progenitor cells than their bone marrow (BM) counterparts, in autologous transplantation protocols. Besides going through a less invasive collection procedure than BM aspiration, leukapheresed CD34+ cell collections ensure a rapid hematologic recovery as a function of transplanted dose of these cells, and their cell cycle status. Patients unable to mobilize sufficient number of CD34+ cells for efficient transplantation procedure are designated as poor mobilizers. Whereas numerous studies were dedicated to defining predictive factors for successful mobilization, only a few characterized the phenotype of mobilized CD34+ in good versus poor mobilizers and none explored the functional and metabolic properties of mobilized cells in these two groups of patients. Thus, Ivanovic et al. (2009) hypothesized that, apart from their mobilization from marrow to the blood, the response to G-CSF of CD34+ cells also includes activation of proliferation, metabolic activity, and proliferative capacity. In this study, mobilized PB CD34+ cells purified from samples obtained by cytapheresis of multiple myeloma or non-Hodgkin's lymphoma patients of both good (>50 CD34+cells/mL) and poor (50 CD34+cells/mL) mobilizers were studied [55]. The initial cell cycle state of CD34+ cells after selection and their kinetics of activation (exit from G0 phase) during ex vivo culture were analyzed. Their proliferative capacity was estimated on the basis of ex vivo generation of total cells, CD34+ cells, and colony-forming cells (CFCs), in a standardized expansion culture. Indirect insight in metabolic activity was obtained on the basis of their survival (viability and apoptosis follow-up) during the 7-day-long conservation in hypothermia (4 °C) in the air or in atmosphere containing 3 % O_2/6 % CO_2. The results have shown that CD34+ cells obtained from good mobilizers were in lower proportion in the G0 phase, their activation in a cytokine-stimulated culture was accelerated, and they exhibited a lower ex vivo expansion efficiency than those from poor mobilizers. The resistance to hypothermia of good mobilizers' CD34+ cells is impaired. The inevitable conclusion was that a good response to G-CSF mobilization treatment is associated with a higher degree of proliferative and metabolic activation of mobilized CD34+cells with a decrease in their expansion capacity [56].

Emphasizing Bioengineering Aspects to Stem Cell Engineering

New concepts in adult stem cell research with development of new strategies: personal experience in the light of significance of growing information.

  Background and significance. Edward Thomas developed bone marrow transplantation as a treatment for leukemia. Initially the process was successful only if the donor was an identical twin of the patient. With the development of immunosuppressant drugs to counter organ rejection now many patients are treated for leukemia, aplastic anemia, sickle cell anemia, Hurler's syndrome, severe combined immunodeficiency (SCID), and *Wiskott-Aldrich syndrome* as a result of his development in bone marrow transplantation. Dr. Edward Thomas was also awarded the *National Medal of Science 1990*. The primary role of adult stem cells in a living organism is to maintain and repair the tissue in which they reside. As an adult, stem cell is an undifferentiated cell found among differentiated cells in a tissue or organ. It can renew itself, and differentiate to yield the major specialized cell types of the tissue or organ. Within past 10 years tremendous piece of work has been done with regard to development of the concepts of "*stemness*," primitive stem cell patterns used in regenerative purposes, and concept of *cancer stem cells*, with significant impact on the development of *new strategies* for their detection and targeted intervention. Despite deep skepticism and arguments these three concepts have their basis in scientific approaches and facts, researched and detected in order to support them. Results obtained are already empowering them to "step" into clinical arena [57–67].

Directions and Relevant Studies: We and Others

What Is "Stemness?"

Stemness has so far been defined as both phenotypically and functionally recognizable cell pattern capable of self-renewal, proliferation, and transdifferentiation through the phenomenon of "plasticity" (Fig. 1.5) [1, 68–72].

Fig. 1.5 Stem cell
maturation according to
D. Zipori. Med Sci (Paris)
2011; 27: 303–301

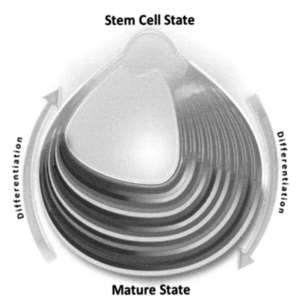

Stem Cell State

Mature State

One has to be aware of the fact that stem cell category, as an elementary term, is assuming the particular functionality. As the entity, or the state, it rationally presents the cell which of its all possible functions possesses at the moment of stemness only those that allow it to survive and sometimes divide; all other functions of this cell are at the potential level. When those possible functions really come up into scenario, that cell is not stem cell anymore. That is why the collections and clusters of different antigens expressed all over the cells in different developmental stages of different tissues (such as kit-receptor, CD117) cannot be the stem cell markers.

The "stemness" is the status in which only the oldest, the most primitive part of the genome is activated with the only purpose to save what is stored in the nucleus of stem cell: genetic information, e.g., potential [1]. The purpose of this event is to save the cell of death and (if it comes to the stimulus for differentiation by asymmetric division) from self-renewal [1]. In that way we are becoming aware that the nature does create the standards that we should rather understand, instead of forcing the nature into our simplified concepts, some of which are very superficial. Tremendous advance which has enabled enrichment of stem cells based upon selection using phenotype as a standard could be appreciated as the advance in this discipline. It has also enabled more direct approach to investigation of stem cells.

However, there are other explanations for this status and one of them was defined by Dove Zipori [68–70]. According to him, this feature is not stem cell specific, given the fact that it is unaquired. Most importantly, according to Zipori, "stemness" is a *transient trait* and cannot be predicted on the basis of momentary gene-expression patterns [71].

We have started optimization of the primitive stem cell pool in the case of acute myocardial infarction with intention to discriminate possible contamination with very small embryonic-like cells (VSELs) within Hematopoietic Stem Cell (HSCs) pool and determine which subpopulation is the best for regenerative purposes.

Optimization of primitive stem cell patterns for regeneration and repair has today at least three strong candidates:

– HSCs (hematopoietic stem cells)
– VSELs (very small embryonic-like stem cells)
– MSCs (mesenchymal stem cells)

The concept of plasticity have been revised by Ratajczak's group which has recently developed and together with us supported the concept of Very Small Embryonic-Like Cells (VSELs), shown to be stem cells in bone marrow and other organs in non-hematopoietic compartment, committed to differentiate into some other tissues. These cells can be detected in mobilized bone marrow cells of mice and humans using cell sorter. However, we have also shown that not all the patients must be good mobilizers which require alternative approach [56]. Therefore, exploring the possibility of using adult stem cells for cell-based therapies has become a very expanding area of investigation (Fig. 1.6).

Breakthrough in stem cell research: John Gurdon (UK) (1933) and Shinya Yamanaka (Japan) (1963) have won the NP in 2012 for the discovery that mature cells can be reprogrammed to become pluripotent

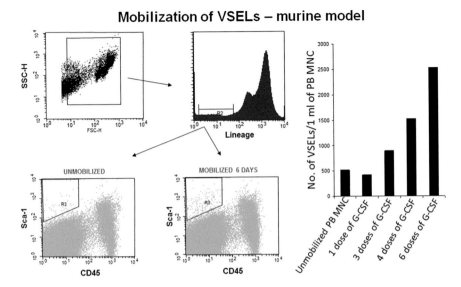

Fig. 1.6 Mobilization of VSELs in murine model by use of G-CSF (Neupogen) in mice and expression of critical markers in mobilized and unmobilized animals. Obtained by kindness of Dr. M. Ratajczak

Breakthrough: Induced Pluripotent Stem Cells

The Yamanaka lab identified four factors that, when co-transfected and expressed in mouse adult fibroblast cells, caused those fibroblasts to revert back to a pluripotent-like state. One year later, the same four factors were used to successfully reprogram the nucleus of human adult fibroblast cells into induced pluripotent stem cells (iPSCs) [73, 74]. These four factors are:

Octamer-4 (Oct-4) encoded by the gene POU5F1 is a transcription factor that is highly expressed in undifferentiated embryonic stem cells compared to other somatic cells. Oct-4 expression in embryonic stem cells is critical to maintain them in an undifferentiated, pluripotent state. In fact, if Oct-4 expression is experimentally knocked out, this causes embryonic stem cells to spontaneously differentiate.

SOX2 is a transcription factor critical for the maintenance of pluripotency in embryonic stem cells. SOX2 and Oct-4 work in parallel to co-regulate expression of target genes involved in the maintenance of pluripotency.

c-Myc is a well-known proto-oncogene. The c-Myc gene codes for a transcription factor that regulates the expression of many genes involved in the control of cell proliferation, growth, differentiation, and apoptosis. Aberrant expression of c-Myc on the other hand is associated with tumor formation and cancer. Recent studies have demonstrated that c-Myc is a dispensable reprogramming factor; however, the transcription factor has been shown to greatly improve reprogramming efficiency.

Kruppel-like factor 4 (Klf-4) is a transcription factor that is highly expressed in undifferentiated ES cells and is also expressed elsewhere in the adult organism including the gut, testis, and lungs and functions to regulate proliferation, differentiation, and cell survival.

Reprogramming as a Therapeutic Event

Recent data have shown the use of reprogramming technologies to cause cancer cells to lose tumorigenicity in chronic myeloid leukemia cells, melanoma cells, and gastrointestinal cancer cells [73–75]. These results suggest that nuclear reprogramming may be a therapeutic strategy for the treatment of cancer [75]. However, these experiments have also revealed that reprogramming technology is not very efficient. Experiments suggest that cancer cells are resistant to reprogramming and this resistance might be related to the role of epigenetic regulations during reprogramming. The fact that transformation of iPSCs is accomplished by erasing the epigenetic modification similar to those found in early embryos demonstrates the significance of epigenetic changes for successful reprogramming, and thus, its role in carcinogenesis [74].

As Dr. Gordana Vunjak-Novakovic published lately, the recent availability of human cardiomyocytes derived from induced pluripotent stem (iPS) cells opens new opportunities to build in vitro models of cardiac disease, screening for new drugs, and patient-specific cardiac therapy [74]. Notably, the use of iPS cells enables studies in the wide pool of genotypes and phenotypes. The progress in reprogramming of iPS cells towards the cardiac lineage/differentiation is going on. The focus is on challenges of cardiac disease modeling using iPS cells and their potential to produce safe, effective and affordable therapies/applications with the emphasis on cardiac tissue engineering. The paper has emphasized implications of human iPS cells to biological research and some of the future needs [74].

Stem cell cancer concept had reached its spike this year [5]. Cancer stem cells (CSCs) share properties similar to those described for tissue stem cells: self-renewal and asymmetric division resulting in the generation of daughter cells destined to differentiate, enabling the regeneration of a tissue. They provide tumors with unrestricted dividing potential that resemble embryonic stem cells. This process of dedifferentiation involves mis-regulation of JAK/STAT, Wnt, and Hedgehog signaling [5]. The theory of cellular dedifferentiation may hold key therapeutic potential in the cancer research field. Proliferation in hypoxic conditions is a common feature of cancer stem cells.

The directions in stem cell research development require further work on:

- More firm and precise definition of stemness.
- Optimization of best candidate for different tissue engineering manipulations in clinical arena and optimization of scaffolds for particular tissue engineered patterns.

- Expanded work on cancer stem cells in order to discriminate origin, underlying causes, and mechanisms of that sort of malignancy and particular, selective, targeted therapeutic approach.

It is clear nowadays that the presence of multipotent stem cells in the adult might open up new therapeutic opportunities on the basis of tissue and organ replacement. Therefore, the exact definition of stem cells and the ability to isolate them are matters of supreme importance. However, despite the efforts of many investigators who strive to determine their nature, a definitive stem cell "portrait" is lacking. Yet, quite recently, two independent studies claimed to have identified a stem cell-specific group of genes that form a "stem cell signature." In fact, these studies have defined two different and unrelated groups of genes; the conclusion that these signatures characterize stem cells is therefore premature. Experimental and/or technical reasons might explain the disparity of the results from these independent studies, and alternative approaches that might lead to identification of the "correct" gene-expression profile of stem cells were suggested. But should one expect to find a stem cell-specific signature using an approach based on the analysis of gene expression? Zipori argues that renewal ability is unaquired property and, as such, is not stem cell specific [70]. Most importantly, according to him, "stemness" is a transient trait and cannot be predicted on the basis of momentary gene-expression patterns. Due to the complexity of the problem, the solution to determining the molecular configurations that dictate a stem cell state should, therefore, come from an overall genomic and proteomic analysis, coupled with mathematical modeling.

Hematopoietic stem cell transplantation remains a risky procedure with many possible complications. It has traditionally been reserved for patients with life-threatening diseases, such as malignancies. While occasionally used experimentally in nonmalignant and nonhematologic indications such as severe disabling autoimmune and cardiovascular diseases, the risk of fatal complications appears too high to gain wider acceptance. Yet, this is the most well-known and the most developed stem cell regenerative approach, given that if successfully engrafted, it repopulates and later on recruits the new, healthy bone marrow cells in circulation.

Embryonic stem cell research is still the matter of controversies at a very stratified levels, although many researchers agree that it might be the source of stem cells with the highest differentiation potential.

The experimental and clinical trials have shown both in animal models and humans the neovascularization and myocardial tissue repair through transdifferentiation into myo-cardiocytes, or some other mechanism. Repair of damaged organ/ tissue (myocardial, neuronal, liver, cartilage, bone, etc.) is shown mostly in animal models, although very good data are coming from the Belgrade group in treatment of AMI [76–81]. Maybe the most illustrative of all is the bunch of experimental data suggesting the great potential for stem cell differentiation and homing into damaged tissues either when mobilized or injected into the tissue of interest after apheresis or BM puncture, with or without cryopreservation [1]. Although the adult stem cell regenerative therapy after BM aspiration and apheresis injection into coronary arteries is becoming more and more successful, the most evident success of mesenchymal stem cell treatment at regenerative therapy level in clinical arena is seen so far in children with *osteogenesis imperfecta* where the results with diseased children dramatically visible and easily reproducible. Yet, due to the obstacles mentioned above, this is not the case with nervous system regenerative treatment, especially in humans.

Apparently, basic adult stem cell research is still evolving, and is the matter of ever-changing issues. Due to our extensive studies, but yet limited knowledge on their behavior and potentials, it is not yet easy to determine how to act in clinical arena. It is obvious that each approach to any particular disease or damage has to be optimized within team work and by bridging the gap between fundamental and clinical studies. Knowing molecular level in depth will help clinicians to orchestrate the team work and overcome critical obstacles in each particular scenario. There is no doubt that adult stem cell therapy (and probably embryonic as well) belong to the future, but we have to act as that we shall belong to the future, as well. Continuous efforts in both molecular and clinical directions will lead to the unique and optimal plan for each particular regenerative treatment. How far away we are from that goal it will inevitably show up in a near future.

References

1. Pavlovic M, Baling B (2013) Stem cells and tissue engineering. Springer, New York. ISBN 978-1-4614-5505-9
2. Saltzman MW (2009) Biomedical engineering. Cambridge University Press, New York. ISBN 978-0-521-840099-6
3. Ratajczak M, Ratajczak J, Shin DM, Wan W, Liu R, Masternak MM, Piotrowska K et al (2011) Higher number of stem cells in the bone marrow of circulating low Igf-1 level Laron Dwarf novel view on Igf-1, stem cells and aging. Leukemia 25:29–733
4. Balint B, Todorović M, Jevtić M, Ostojić G, Ristanović E, Vojvodić D et al (2009) The use of stem cells for marrow repopulation and in the field of regenerative medicine. Makedon Med Pregl 63(Suppl 75):12–17
5. Mayfield J, Pavlovic M (2014) Current modalities and the implications of cancer stem cell engineering in oncological treatment. ART 14(1–2):49–59
6. Pavlovic M, Balint B (2006) The use of stem cells to repair the cardiac tissue. Anest Reanim Transfuziol 34:129–150
7. Li C et al (2007) Identification of pancreatic cancer stem cells. Cancer Res 67(3):1030–1037
8. Orlic D, Kajstura J, Chimenti S et al (2001) Bone marrow cells regenerate infarcted myocardium. Nature 410(6829):701–705
9. Zhang M, Methot D, Poppa V, Fujio Y, Walsh K, Murry C (2001) Cardiomyocyte grafting for cardiac repair: graft cell death and anti-death strategies. J Mol Cell Cardiol 33(5):907–921
10. Dow J, Simkhovich BZ, Kedes L, Kloner RA (2005) Washout of transplanted cells from the heart: a potential new hurdle for cell transplantation therapy. Cardiovasc Res 67(2):301–307
11. http://www.ncbi.nlm.nih.gov/pubmed/17645413
12. http://www.intechopen.com/books/apoptosis-and-medicine/apoptosis-as-a-therapeutic-target-in-cancer-and-cancer-stem-cells-novel-strategies-and-futures-persp
13. http://stemcells.nih.gov/info/Regenerative_Medicine/pages/2006chapter9.aspx
14. Neve RM et al (2006) A collection of breast cancer cell lines for the study of functionally distinct cancer subtypes. Cancer Cell 10(6):515–527
15. Al-Hajj M et al (2003) Prospective identification of tumorigenic breast cancer cells. Proc Natl Acad Sci U S A 100(7):3983–3988

16. Reya T et al (2001) Stem cells, cancer, and cancer stem cells. Nature 414(6859):105–111
17. Singh SK et al (2003) Identification of a cancer stem cell in human brain tumors. Cancer Res 63(18):5821–5828
18. Kucia M, Wojakowski W, Reca R, Machalinski B, Gozdzik J, Majka M, Baran J, Ratajczak J, Ratajczak M (2006) The migration of bone marrow-derived non-hematopoietic tissue committed stem cells is regulated in an SDF-1, HGF—LIF-dependent manner. Arch Immunol Ther Exp (Warsz) 54(2):121–35
19. Pavlović M (2008) VSELs concept: review. MNE Medica 1:16–17, ref. p. 43
20. Pavlovic M (2014) Very small embryonic like cells (VSELs): review and perspectives in the light of critical data and controversies. ART 41(1–2):33–47
21. Ratajczak MZ, Kucia M, Reca R, Majka M et al (2004) Stem cell plasticity revised: CXR4 positive cells expressing mRNA for early muscle, liver and neural cells "hide out" in the bone marrow. Leukemia 19(1):29–40
22. Kucia M, Reca R, Campbell FR, Zuba-Surma E, Majka M, Ratajczak J, Ratajczak MZ (2006) A population of very small embryonic-like (VSEL) CXCR4þSSEA-1þOct-4þ stem cells identified in adult bone marrow. Leukemia 20:857–869
23. Kucia M, Halasa M, Wysoczynski M, Baskiewicz-Masiuk M, Moldenhawer S, Zuba-Surma E, Czajka R, Wojakowski W, Machalinski B, Ratajczak MZ (2007) Morphological and molecular characterization of novel population of CXCR4þ SSEA-4þOct-4þ very small embryonic-like cells purified from human cord blood—preliminary report. Leukemia 21:297–303
24. Zuba-Surma E, Kucia M, Abdel-Latif A, Dawn B, Hall B, Singh R, Lillard JW, Ratajczak MZ (2008) Morphological characterization of very small embryonic-like stem cells (VSELs) by image stream system analysis. J Cell Mol Med 12(1):292–303
25. Zuba-Surma EK, Wu W, Ratajczak J, Kucia M, Ratajczak MZ (2009) Very small embryonic-like stem cells in adult tissues-potential implications for ageing. Mech Ageing Dev 130(1–2):58–66
26. Ratajczak M, Zuba-Surma E, Machslinski B, Ratajczak J, Kucia M (2008) Very small embryonic-like (VSEL) stem cells: purification from adult organs, characterization, and biological significance. Stem Cell Rev 4(2):89–99
27. Kucia M, Wysoczynski M, Ratajczak J, Ratajczak MZ (2008) Identification of very small embryonic like (VSEL) stem cells in bone marrow. Cell Tissue Res 331:125–134
28. Suszynska M, Zuba-Surma EK, Maj M, Mierzejewska K, Ratajczak J, Kucia M, Ratajczak MZ (2014) The proper criteria for identification and sorting of very small embryonic-like stem cells, and some nomenclature issues. Stem Cells Dev 23(7):702–13
29. Ratajczak M, Shin DM, Liu R et al (2012) Very small embryonic/epiblast like stem cells (VSELs) and their potential role in aging and organ rejuvenation-an update and comparison to other primitive small stem cells isolated from adult tissues. Aging (Albany NY) 4(4):235–46
30. Rodgerson DO, Ratajczak M. http://www.neostem.com/assets/ Where do stem cells come from. Pdf (Rodgerson DO, Harris AG (2011) A comparison of stem cells for therapeutic use. Stem Cell Rev 7(4):782–796
31. Kassmer SH, Bruscia EM, Zhang PX, Krause DS (2012) Nonhematopoietic cells are the primary source of bone marrow-derived lung epithelial cells. Stem Cells 30(3):491–499
32. Kassmer SH, Krause DS (2013) Very small embryonic-like cells: biology and function of these potential endogenous pluripotent stem cells in adult tissues. Mol Reprod Dev 80:677–690
33. Kassmer SH, Jin H, Zhang P-X, Bruscia EM, Heydari K, Lee JH, Kim CF, Kassmer SH (2013) Very small embryonic-like stem cells from the murine bone marrow differentiate into epithelial cells of the lung. Stem Cells 31(12):2759–2766
34. Bhartiya D (2013) Are mesenchymal cells indeed pluripotent stem cells or just stromal cells? OCT-4 and VSELs biology has led to better understanding. Stem Cells Int 2013, 547501, Epub 2013 Sep 25. Review
35. Anand S, Bhartiya D, Sriraman K, Patel H, Manjramkar D, Bakshi G, Dhamankar V, Kurkure P (2013) Quiescent very small embryonic-like stem cells resist oncotherapy and can restore spermatogenesis in germ cell depleted mammalian testis. Stem Cells Dev. doi:10.1089/scd.2013.0059

36. Bhartiya D, Sriraman K, Parte S, Patel HJ (2013) Ovarian stem cells: absence of evidence is not evidence of absence. Ovarian Res 6(1):65. doi:10.1186/1757-2215-6-65.13

37. Bhartiya D, Unni S, Parte S, Anand S (2013) Very small embryonic-like stem cells: implications in reproductive biology. Biomed Res Int 2013, 682326. doi:10.1155/2013/682326

38. Bhartiya D, Kasiviswananthan S, Shaikh A (2012) Cellular origin of testis-derived pluripotent stem cells: a case for very small embryonic-like stem cells. Stem Cells Dev 21(5):670–674

39. Bhartiya D, Shaikh A, Nagvenkar P, Kasiviswanathan S, Pethe P, Pawani H, Mohanty S, Rao SG, Zaveri K, Hinduja I (2012) Very small embryonic-like stem cells with maximum regenerative potential get discarded during cord blood banking and bone marrow processing for autologous stem cell therapy. Stem Cells Dev 21(1):1–6

40. Wang J, Guo X, Lui M, Chu PJ, Yoo J, Chang M, Yen Y (2014) Identification of a distinct small cell population from human bone marrow reveals its multipotency *in vivo* and *in vitro*. PLoS One 9(1):e85112. doi:10.1371/journal.pone.0085112

41. Wojakowski W, Tendera M, Kucia M, Zuba-Surma E, Paczkowska E, Ciosek J, Halasa M, Krol M, Kazmierski M et al (2009) Mobilization of bone marrow-derived Oct-4+SSEA-4+very small embryonic-like stem cells in patients with acute myocardial infarction. J Am Coll Cardiol 53:1–9

42. Chang YJ, Tien KE, Wen CH, Hsieh TB, Hwang SM (2014) Recovery of CD45-/Lin-/SSEA-4+very small embryonic-like stem cells by cord blood bank standard operating procedures. Cytotherapy 16(4):560–5. doi:10.1016/j.jcyt.2013.10.009

43. Knoepfler P (2013) Stem cell blog. Taichman are VSELs the sasquatch of the stem cell field? http://www.ipscell.com/tag/russell-taichman

44. Szade K, Bukowska-Strakova K, Nowak WN, Szade A, Kachamakova-Trojanowska N, Zukowska M, Jozkowicz A, Dulak J (2013) Murine bone marrow lin–Sca-1+CD45– very small embryonic-like (VSEL) cells are heterogeneous population lacking Oct-4A expression. PLoS One 8(5), e63329. doi:10.1371/journal.pone.0063329

45. Miyanishi M, Mori Y, Seita J, Chen JY, Karten S, Chan CK, Nakauchi H, Weissman IL (2013) Do pluripotent stem cells exist in adult mice as very small embryonic stem cells? Stem Cell Reports 1(2):198–208

46. D'Ipolito G, Diabira S, Howard GA, Menei P, Roos BA, Schiler PC (2004) Marrow–isolated adult, multilineage inducible (MIAMI) cells, a unique population of postnatal and old human cells with extensive expansion and differentiation potential. J Cell Sci 117:2971–2981

47. Jiang Y, Jahagirdar BN, Reinhardt RL, Schwartz RE, Keene CD, Ortiz-Gonzalez XR, Reyes M, Lenvik T, Lund T, Blackstad M, Du J, Aldrich S, Lisberg A, Low WC, Largaespada DA, Verfaillie CM (2002) Pluripotency of mesenchymal stem cells derived from adult marrow. Nature 418(6893):41–49

48. Yoon YS, Wecker A, Heyd L, Park JS, Tkebuchava T, Kusano K, Hanley A, Scadova H, Asahara T, Losordo DW (2005) A Clonally expanded novel population of multipotent stem cells derived from human bone marrow regenerates myocardium after myocardial infarction. J Clin Invest 115:326–338

49. Erices A, Conget P, Minguell JJ (2000) Mesenchymal progenitor cells in human umbilical cord blood. Br J Haematol 109(1):235–242

50. Huss R et al (2000) Perspectives on the morphology and biology of CD34-negative stem cells. J Hematother Stem Cell Res 9(6):783–793

51. Zvaifler NJ, Marinova-Mutafchieva L, Adams G, Edwards CJ, Moss J, Burger JA, Maini RN (2000) Mesenchymal precursor cells in the blood of normal individuals. Arthritis Res 2(6):477–488

52. Young HE, Steele TA, Bray RA, Hudson J, Floyd JA, Hawkins K, Thomas K, Austin T, Edwards C, Cuzzourt J, Duenzl M, Lucas PA, Black AC Jr (2001) Human reserve pluripotent mesenchymal stem cells are present in the connective tissues of skeletal muscle and dermis derived from fetal, adult, and geriatric donors. Anat Rec 264(1):51–62

53. Ratajczak J, Kucia M, Reca R, Zhang J, Machalinski B, Ratajczak MZ (2003) Quiescent CD34+ early erythroid progenitors are resistant to several erythropoietic "inhibitory" cytokines; role of FLIP. Br J Haematol 123(1):160–169

54. Zuk PA, Zhu M, Mizuno H, Huang J, Futrell JW, Katz AJ, Benhaim P, Lorenz HP, Hedrick MH (2001) Multilineage cells from human adipose tissue: implications for cell-based therapies. Tissue Eng 7(2):211–228

55. Ivanovic Z, Kovacevic-Filipovic M, Jeanne M, Ardilouze L, Bertot A, Szyporta M, Hermitte F, Lafarge X, Duchez P, Vlaski M, Milpied N, Pavlovic M, Praloran V, Boiron JM (2010) CD34+ cells obtained from "good mobilizers" are more activated and exhibit lower *ex vivo* expansion efficiency than their counterparts from "poor mobilizers". Transfusion 50(1):120–127

56. Kucia MJ, Wysoczynski M, Wu W (2008) Evidence that very small embryonic-like stem cells are mobilized into peripheral blood. Stem Cells 26:2083–2092

57. Korbling M, Katz RL, Khanna A, Ruifrok AC, Rondon G, Albitar M, Champlin RE, Estrov Z (2002) Hepatocytes and epithelial cells of donor origin in recipients of peripheral-blood stem cells. New Engl J Med 346:738–746

58. Kucia M, Masternak M, Liu R, Shin DM, Ratajczak J, Mierzejewska K et al (2013) The negative effect of prolonged somatotrophic/insulin signaling on an adult bone marrow-residing population of pluripotent very small embryonic-like stem cells (VSELs). Age (Dordr) 35:315–330

59. Kucia M, Shin DM, Liu R, Ratajczak J, Bryndza E, Masternak MM et al (2011) Reduced number of VSELs in the bone marrow of growth hormone transgenic mice indicates that chronically elevated Igf1 level accelerates age-dependent exhaustion of pluripotent stem cell pool: a novel view on aging. Leukemia 25:1370–1374

60. Ratajczak MZ, Shin DM, Kucia M (2009) Very small embryonic/epiblast-like stem cells: a missing link to support the germ line hypothesis of cancer development? Am J Pathol 174(6):1985–1992

61. Shin DM, Liu R, Klich I, Ratajczak J, Kucia M, Ratajczak MZ (2010) Molecular characterization of isolated from murine adult tissues very small embryonic/epiblast like stem cells (VSELs). Mol Cells 29(6):533–538

62. Ratajczak MZ, Liu R, Ratajczak J, Kucia M, Shin DM (2011) The role of pluripotent embryonic-like stem cells residing in adult tissues in regeneration and longevity. Differentiation 81(3):153–161

63. Ratajczak MZ, Liu R, Marlicz W, Blogowski W, Starzynska T, Wojakowski W, Zuba-Surma E (2011) Identification of very small embryonic/epiblast-like stem cells (VSELs) circulating in peripheral blood during organ/tissue injuries. Methods Cell Biol 103:31–54

64. Ratajczak MZ, Zuba-Surma EK, Shin DM, Ratajczak J, Kucia M (2008) Very small embryonic-like (VSEL) stem cells in adult organs and their potential role in rejuvenation of tissues and longevity. Exp Gerontol 43(11):1009–1017

65. Ratajczak MZ, Zuba-Surma E, Wojakowski W, Synska M, Mierzejewska K, Liu R, Ratajczak J, Shin M, Kucia M (2013) Very small embryonic-like stem cells (VSELs) represent a real challenge in stem cell biology: recent pros and cons in the midst of a lively debate. Leukemia 28(3):473–483

66. Suszynska M, Zuba-Surma EK, Maj M, Mierzejewska K, Ratajczak J, Kucia M, Ratajczak MZ (2014) The proper criteria for identification and sorting of very small embryonic-like stem cells, and some nomenclature issues. Stem Cells Develop 23:1–13

67. Havens AH, Shiozawa SY, Jung Y, Wang J, Mishra A, Jiang Y, O'Neill DW, Krebsbach PH et al (2013) Human and murine very small embryonic-like (VSEL) cells represent multipotent tissue progenitors, *in vitro* and *in vivo*. Stem Cells Dev 23:689–701. doi:10.1089/scd.2013.0362

68. Zipori D (2004) The nature of stem cells: state rather than entity. Nat Rev Genet 5(11):873–878

69. Zipori D (2009) Biology of stem cells and the molecular basis of the stem state. Humana, New York

70. Zipori D (2011) À la recherche d'une définition moléculaire plus que descriptive pour les cellules souches towards a molecular rather than a descriptive definition of stemness. Med Sci (Paris) 27:301–303

71. Ramalho-Santos M et al (2002) Stemness: transcriptional profiling of embryonic and adult stem cells. Science 298(5593):597–600

72. http://www.nature.com/nature/focus/cancerstemcells/
73. Martins AM, Vunjak-Novakovic G, Reis RL (2014) The current status of iPS cells in cardiac research and their potential for tissue engineering and regenerative medicine. Stem cell Rev 10:177–190
74. Yamanaka S (2010) Patient-specific pluripotent stem cells become even more accessible. Cell Stem Cell 7(1):1–2
75. Novakovic-Vunjak G, Scaden DT (2011) Biomimetic platforms for human stem cell research. Cell Stem Cell 8(3):252–261
76. Balint B, Stamatović D, Todorović M, Jevtić M, Ostojić G, Pavlović M, Lojpur Z, Jocić M (2007) Stem cells in the arrangement of bone marrow repopulation and regenerative medicine. Vojnosanit Pregl 64(7):481–484
77. Obradovic S, Rusovic S, Dincic D, Gligic B, Baskot B, Balint B et al (2003) Autologous pluripotent progenitor cells in the treatment of ischemic heart disease. Vojnosanit Pregl 60(6):725–731
78. Venkatraman AX, He C, Thorvaldsen JL, Sugimura R, Perry JM, Tao F, Zhao M, Christenson MK, Sanchez R et al (2013) Maternal imprinting at the H19-Igf2 locus maintains adult haematopoietic stem cell quiescence. Nature 500:345–349
79. Balint B, Ivanovic Z, Petakov M, Taseski J, Jovcic G, Stojanovic N et al (1999) The cryopreservation protocol optimal for progenitor recovery is not optimal for preservation of MRA. Bone Marrow Transplant 23:613–619
80. Balint B, Jovicic D, Todorovic M, Subota V, Pavlovic M, Goodrich R (2013) Plasma constituent integrity in pre-storage vs. post-storage riboflavin and UV–light treatment—a comparative study. Transfus Apher Sci 49:434–439
81. Balint B, Pavlovic M, Todorovic M, Jevtic M, Ristanovic E, Ignjatovic L (2010) The use of simplified *ex vivo* immunoadsorption and "multi–manner" apheresis in ABO/H-mismatched kidney transplant setting—phase II clinical study. Transfus Apher Sci 43(2):141–148

Chapter 2
Normal Stem Cells: Biology, Collection/ Harvesting, and Ex Vivo Manipulations

The good thing about science is that it's true whether or not you believe in it.

Neil deGrasse Tyson

Introduction

Stem cells (SCs) have an extensive self-renewal capacity and a high potential for proliferation as well as differentiation into pluripotent or committed progenitors and mature blood cells. Hematopoietic events with a complex network of interactive cytokines (grow factors and their inhibitors) are regulated. Different population of SCs expresses CD34 antigen, consequently they are named also as CD34+ cells [1–4]. Thanks to abovementioned characteristics, SCs provide complete and long-term bone marrow (BM) repopulation with subsequent hematopoietic reconstitution after transplantation. A traditional SC transplantation involves myelo (immuno) ablation—the administration of intensive radio-chemotherapy—followed by (re) infusion of harvested cells in order to eliminate of basic disease [5–7]. Similar procedure with reduced-intensity conditioning (RIC) can be offered to patients who are disqualified for high-dose radio-chemotherapy because of their age or comorbidity [8]. Malignant disorders are the most common indication for this therapeutic modality [9–12]. SC transplantation is also used for therapy of benign diseases, such as severe combined immunodeficiency (SCID), metabolic or autoimmune disorders [13–15].

In a few words, in different clinical settings totipotent, pluripotent, and multipotent SCs give rise to repopulation of recipient's BM (engraftment) with subsequent complete, stable, and long-term reconstitution of hematopoiesis. In addition, they are also capable of colonizing different tissues ("homing"). Thus, initial experimental and clinical studies showed that therapeutic use or "implantation" of autologous SCs into damaged and/or ischemic area induces their "homing" and following "transdifferentiation" into the cell lineages of host organ, including collateral vessel formation. Angiogenic growth factors (or genes encoding of these proteins) promote the development of collateral arterioles, and the process is called as "therapeutic angiogenesis" or "neovascularization" [1–3].

© The Author(s) 2015
M. Pavlovic, B. Balint, *Bioengineering and Cancer Stem Cell Concept*,
DOI 10.1007/978-3-319-25670-2_2

In practice, SCs could be collected by multiple aspirations from BM, by mononuclear cell (MNC) harvesting from PB (after mobilizing regimen) or by purification from umbilical cord blood (UCB). Typically, the use of BM or PB derived grafts (allogeneic or autologous) is a standard method in adult setting. UCB transplants have provided hopeful results firstly in pediatric patients—when a matched unrelated BM or PB donor is unavailable [4, 16–25]. In relation to genetic, and particularly HLA relationship between donor and recipient, transplantations can be classified as autologous, allogeneic and syngeneic [1, 4]. The use of autologous transplantation requires both optimized harvesting procedures to get enough SC yield and cryopreservation that guarantees the best possible cell recovery. Despite the fact that cryopreservation are already in routine use, some questions related to optimal freezing method and cryoprotectant (e.g., dimethyl sulfoxide—DMSO, hydroxyetilstarch—HES) type and concentration are not resolved [26–31]. For marrow failure (immunodeficiency, severe aplastic anemia (SAA), BM infiltration) and metabolic disorders, the use of allotransplantation is the therapeutic method of choice if patient has HLA-matched donor. However, allogeneic SC application is not without a risks, such as graft failure, despite intensive myeloablative conditioning regimen applied as well as acute and/or chronic Graft versus Host Disease (GvHD), regardless of immunosuppressive therapy administered [4, 32–34].

Generally, SCs can be divided into embryonic and "tissue-specific" (adult) cell compartment [2]. Current researches have recognized that some adult SCs have similar "unlimited" biological potential than embryonic cells. Consequently, adult SCs are able to develop into a variety of somatic cells by "transdifferentiation" or "SC plasticity" [35–37]. Although the term "cell plasticity" became very popular, some studies have suggested that BM might contain different types of SCs that can produce non-hematopoietic (somatic) cells. For example, mesenchymal SCs in BM give rise to osteocytes, chondrocytes, adipocytes, and skeletal muscle. Consequently, adult BM or peripheral blood (PB) derived SCs are clinically applicable for the cell therapy in the field of regenerative medicine, that is cell/organ replacement and/or regeneration—precisely for the treatment of patients with myocardial, liver, vascular, neurological, or other tissue damages [38–43]. The SC applications used in the Institute of Transfusiology and Hemobiology of MMA are presented in Fig. 2.1.

Autologous SC transplants. When is BM appropriate to use as source of SCs for autologous resuscitation following myeloablative radio-chemotherapy or RIC depends on the marrow general state and/or infiltration with malignant cells. Fibrosis makes marrow not possible for SC collection by aspirations. Tumor cell infiltrates eliminate marrow as a transplant source as well. Prior pelvic irradiation, poor anesthesia risk, obesity, or patient refusal of marrow collection can limit marrow as an option. Mobilized autologous SCs from PB are commonly used in the above situations and in heavily treated patients. This procedure is nowadays in routine clinical practice and provides more progenitor cell yield than conventional marrow harvest and therefore earlier engraftment, which is a faster hematopoietic recovery. Primarily for this reason, transplant of PB derived SCs has practically replaced BM

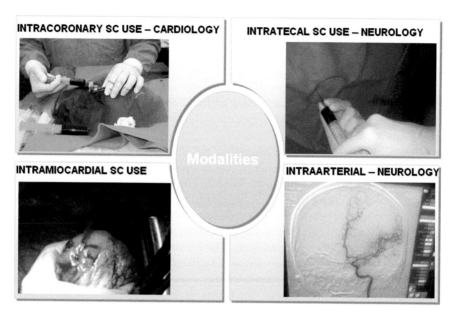

INTRACORONARY SC USE – CARDIOLOGY

INTRATECAL SC USE – NEUROLOGY

INTRAMIOCARDIAL SC USE

INTRAARTERIAL – NEUROLOGY

Modalities

Fig. 2.1 Stem cell application approaches in our center for regenerative medicine

transplant in an autologous setting. As mentioned, autologous PB harvests involve mobilizing the SCs from the patient's BM compartment into the circulation using different growth factors, typically in combination with chemotherapy prior to collection. Once in the circulation, the SCs are collected by apheresis—conventional or large-volume leukapheresis [1–3, 6, 44].

Allogeneic SC transplants. Transplant of allogeneic SCs are indicated in the treatment of patients with malignant disease—if they have HLA-matched donors. For patients with immunodeficiency, marrow failure, metabolism disorders, etc., the use of allogeneic SCs is imperative. However, there are also some "atypical data" related to treatment of SAA using autologous SCs [10]. Allogeneic transplant is associated with a risk that immunocompetent donor cells will react against recipient tissues (GvHD), despite immunosuppressive therapy administered. In adult "related allogeneic setting," the best results are obtained using completely HLA-matched (HLA-identical, i.e., six-antigen-matched donor/recipient pairs) transplants. There is a 25 % chance of a sibling being a complete match, a 50 % chance of a haplotype match, and a 25 % chance of a complete mismatch. Pediatric patients are more tolerant of partially mismatched graft [1–4, 44, 45].

Data obtained up till now has shown that the use of SC donor registers can successfully recruit unrelated donors for collection of BM or PB derived SCs. Thus, matched unrelated donor searches can be initiated for approximately 70 % of candidates without sibling donor. These protocols have possible benefit since higher engraftment potential of allogeneic vs. autologous SCs and following earlier hematopoietic reconstitution, as well as occurrence of GvL effect [1, 3, 44]. For definitive choice, additional experimental and clinical trials for comparison of efficacy and outcome of autologous vs. allogeneic (related or unrelated) BM vs. PB derived SC transplant are required.

Syngeneic and haploidentical SC transplants. Occasionally recipient has an identical twin—a syngeneic transplant is optimal because the donor and recipient cells are genotypically identical (the first transplants performed in humans) [39]. On the other hand, syngeneic grafts do not induce graft vs. tumor that is GvL effect in recipient. Our knowledge of the immunobiology of SC transplant across major histocompatibility complex (MHC) barriers—haploidentical transplants—has increased significantly over the past decades. The key reason (or limitation) for realization potential haploidentical SC transplant is the absence of a HLA-matched related donor in the majority of families. On the other hand, the conversion of a new hypothetical therapeutic option into the routine haploidentical SC transplant clinical practice is accepted and developed more slowly. The most critical complications of SC transplants across HLA barriers are the graft rejection and/or occurrence of GvHD. However, these adverse events maybe could be successfully prevented and treated using current pharmacologic approaches or manipulations during the haploidentical hematopoietic grafting process [1].

In a few words, the intensifying of myelo (immuno) ablative therapy combined with SC transplantation and the introduction of cell-mediated restorative/regenerative methods ("cell therapy") resulted in increased needs for both SCs conceptual and practical operating procedures inducing minimized cell damages during their harvesting and cryopreservation. In this article, data in the field of practical aspects of an optimized SC harvesting, purification, and cryopreservation will be briefly reviewed. In addition, our results of the investigation of SC harvesting and different graft ex vivo "graft engineering" (cell processing, selection and cryopreservation) protocols will be summarized.

Bone Marrow Derived Stem Cell Collection

Historically, BM was the first source of SCs for transplant in experimental and clinical setting [4–7]. A marrow harvest is the same for an allogeneic donor as for an autologous patient. SCs are collected by multiple aspirations from the posterior and anterior iliac crest and (seldom) from sternum. The posterior iliac crest provides the richest site of marrow. The procedure is performed under sterile conditions in the operation room, while the donor is generally anesthetized (Fig. 2.2).

In order to provide required number of nucleated cells (TNCs), that is $\geq 3 \times 10^8/$ kg of body mass (kgbm), around 200 aspirations are required, where single aspirate volume is 2–5 mL. Immediately after the collection, cell aspirate should be filtered in order to remove bone and lipid tissue particles and/or cell aggregates. Anticoagulation is provided using solution containing citrate and by heparin diluted

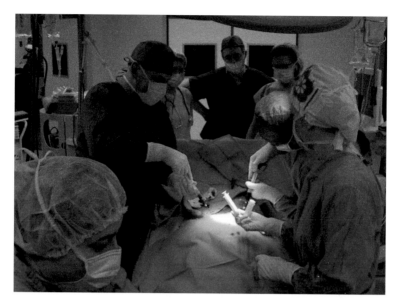

Fig. 2.2 BM derived SC collection by multiple aspirations

in saline (5000 IU/500 mL), using autologous plasma or one of the cell culture medium for resuspension of collected cells [1–7, 44, 45].

The target dose of collected marrow is 10–15 mL per kgbm. Thus, the volume of aspirate is relatively large (800–1000 mL) and it contains a high count of red blood cells. Accordingly, in order to prevent anemia in donors, blood for autologous transfusion (to carry out during SC collection) should be collected around 1 week before SC collection and transplant [1–3, 44].

On the other hand, of aspirate volume, precisely red blood cell number and/or plasma quantity reduction is required (by processing—Fig. 2.3), especially for ABO incompatible (major and/or minor) transplants or when cryopreservation is intended (autologous setting).

A commonly used minimum target (after processing) of TNC count—for both autologous and allogeneic transplants—is 2×10^8/kgbm [1–4, 44]. The concentration of the CD34$^+$ cells and/or depletion of T-cells (positive/negative cell selection) in final cell unit is achieved by the ex vivo purging procedure using immunomagnetic device for cell selection [2, 44].

These SC purification procedures (processing and purging or selection) enable reduction of the aspirate volume, i.e., reduction of red blood cell for around 80–90 %, or even more precisely, the depletion of mentioned unwanted (malignant or T-cells) cells with efficacy ≥3–4 Log$_{10}$ [1–4, 45]. Development of the ability to isolate selected SCs and/or ex vivo expand them into a large number is expected to broad their beneficial therapeutic effects, since the limitation to many of SC applications has been the absolute number of defined target cells.

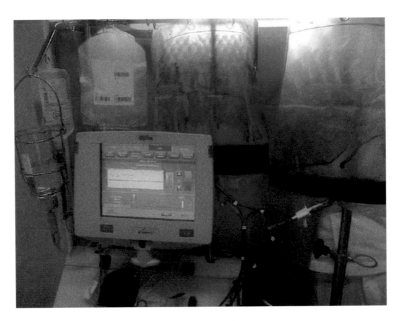

Fig. 2.3 BM aspirate processing after SC collection using Spectra-Optia device

Peripheral Blood Derived Stem Cell Harvesting/Purification

CD34 is the cluster designation given to a transmembrane glycoprotein present on SC surface and some stromal cells. Cells expressing the CD34 antigen (obtained from BM or PB) are capable of complete reconstitution of hematopoiesis. The first SC harvests from PB were accomplished in "steady state hematopoiesis"—but using numerous [6–9] collections and following cryopreservation was needed [1–3]. Currently, SCs are harvested after mobilization by the use of chemotherapy and/ or recombinant colony-stimulating factors (rHuG-CSF). The typical number of apheresis required is not more than 1–3.

The collection PB-SCs is an aphaeretic procedure with respect to the standardized protocol and cell yield (Fig. 2.4).

Characteristically, for anticoagulation an acid-citrate-dextrose formula B (ACD-B with 1.8 % citrate concentration) or ACD-A (2.2 % citrate concentration) solutions are used alone (seldom in combination with heparin). For allogeneic transplantations, venous access is most frequently realized through antecubital veins. In autologous setting, collection should be performed across central-venous, jugular, or femoral vascular access. Short-term use of femoral catheters appears safe and effective, improving patient comfort and reducing cost. These catheters have simplified cell harvesting, but may be associated with thrombosis of the instrumented vessels. In addition, there is approximately one percent central-venous catheter-related hazard of the local infection, pneumothorax or bleeding [1–3, 18–22, 44].

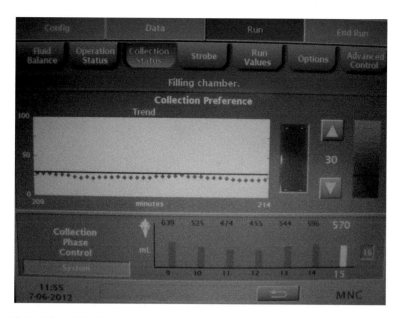

Fig. 2.4 Peripheral blood SC harvesting and monitoring by Spectra-Optia device

PB-SC transplantation could be described by: (a) absence of general anesthesia and work in surgical division; (b) smaller harvest quantity and higher cell yield; (c) improved engraftment rate and lower transplantation-related morbidity; (d) earlier/faster hematopoietic and immune reconstitution. Due to the mentioned reasons, the number of patients treated by PB-SCs is ever increasing worldwide, especially in autologous SC transplantation setting [1, 44].

For obtaining acceptable SC or CD34+ yield, efficient mobilization protocol is required. Allogeneic donors are given rHuG-CSF 5–10 µg/kgbm daily subcutaneously. The CD34+ cell count in the circulation begins to rise after 2–3 days of rHuG-CSF administration and peaks is on the fifth day. When donor mobilization with rHuG-CSF is poor, the only ways to improve yields are to increase the blood volume processed or the number of collections. In autologous setting typical rHuG-CSF doses are higher—patients are given rHuG-CSF 12–16 µg/kgbm daily combined with mono-chemotherapy (cyclophosphamide 4–7 g/m²) or by poly-chemotherapy in corresponding doses [1, 18–20].

In the course of cell harvesting, the determination of the optimized collection system (Fig. 2.5) and optimal timing for apheresis are the most critical event.

For allogeneic donors the first apheresis is on the fifth day of rHuG-CSF application. However, the definition of best possible timing of autologous collection from patients who primed by chemotherapy plus rHuG-CSF is more complex and controversial. The optimal timing can be determined based on the specific cell values in the hemogram. Leukocyte and/or MNC counts, as well as the number of circulating CD34+ cells, have all been used as markers to determine when to initiate harvesting.

Fig. 2.5 The most recent peripheral blood SC harvesting approach—the use of the Intermediate density layer (IDL) system

It is suggested that optimal time to begin cell collection is when the leukocyte count $\cong 5$–10×10^9/L. However, the leukocytes do not correlate strongly with the number of SC in the graft. Contrary, circulating CD34$^+$ number evidently correlates with collection timing and the SC quantity in harvest (as a function of the volume of blood processed also). Namely, it is presented that for a CD34$^+ \geq 20$–40/μL of patient's blood the possibility of the CD34$^+$ yield $\geq 2.5 \times 10^6$ cells/kgbm is around 15 % after performance of one "standard" collection or 60 % or more after one LVL. Of course, higher CD34$^+$ number in circulation results in superior yield [1, 17–19]. Our results also confirmed high-level efficacy of the LVL. Namely, for the 89.5 % patients using one LVL, the mean CD34$^+$ yield was 12.1×10^6/kgbm (allogeneic) and 6.5×10^6/kgbm (autologous), respectively. In our group of patients, the circulating CD34$^+$ count was also relatively high 40–60/μL following mobilizing regiment [1, 44, 45].

The efficiency and standardization of PB-SC harvesting can be estimated by MNC (now seldom) and CD34$^+$ (typically) quantifications. It is generally considered that the count of MNC should be about 300×10^8 per unit, that is ≥ 2–4×10^8/kgbm, and CD34$^+$ around 330×10^6 per unit, that is ≥ 2–4×10^6/kgbm in harvest (contamination: no more than 470×10^9 platelets and 7.6 mL red blood cell total volume per unit) in order to expect successful transplant. However, recent data support a clinical benefit associated with greater CD34$^+$ yield ($\geq 5.0 \times 10^6$/kgbm) compared to the minimum cell quantity needed ($\geq 1.0 \times 10^6$/kgbm) in autologous setting [1, 27, 28]. Although commonly accepted, the stated cell yields cannot guarantee stable and long-term reconstruction of hematopoiesis following transplant. In order to achieve them, the next details are needed: (a) the volume of processed blood in one apheresis ≥2–3 or more (intensive or LVL) patient's circulating volume, that is around 16–25 L for person with around 70–80 kg of body mass and (b) apheretic procedures should be performed 1–2 (occasionally more) times [1, 20, 27, 44].

However, patients who have been earlier treated with high-dose radio-chemotherapy may be "poor responders" for chemotherapy plus rHuG-CSF induced mobilization. The most efficient approach to obtain adequate SCs from "poor mobilizers" is not resolute still. Simultaneously collection of SCs from BM and PB has not improved engraftment rate significantly. Mobilization with rHuG-CSF alone is perhaps more efficient than rHuG-CSF in addition to chemotherapy. Increased doses of rHuG-CSF or use of rHuG-CSF together with rHuGM-CSF has also effectively mobilized some autologous donors [1, 45].

When PB-SCs are collected, additional harvest ex vivo manipulation, that is, processing (to obtain a red blood cell count and/or plasma volume reduction), is not required because the final hematocrit is small (between 0.05 and 0.10) and the final volume of cell suspension is only around 200–250 mL [44]. On the other hand, the main disadvantage of the use of PB-SCs is high-level T-lymphocyte and occasionally tumor cell "contamination" and subsequent risk of GvHD or disease relapse. However, efficient graft purification methods, that is, depletion of tumor cells (autologous) or T-lymphocytes (allogeneic), were developed using immunomagnetic technique (positive and/or negative selection) by cell sorters [2, 11, 12]. Namely, earlier studies reported an elevated risk of GvHD in allogeneic PB-SC vs. BM-SC recipients, but recent prospective randomized studies found no differences in the incidence of this complication [1, 44]. The use of stated graft purification by immunomagnetic system has been shown to be the most effective method to achieve a 3–4 Log_{10} T-lymphocyte depletion while retaining around 60–80 % of the $CD34^+$ cells in the graft [1, 11, 12].

Our results (Fig. 2.6) also verified that $CD34^+$ cell recovery was 70–80 %, when post-selection $CD34^+$ purity ($CD34^+$ cell percentage in final cell suspension) was around 80–90 % [1–3, 36, 44].

In addition, the objective of our preclinical researches during last years was to optimize SC collection and processing protocols, as well as mobilization/harvesting timing in order to obtain high $CD34^+$ and especially a more primitive $CD34^+/CD90^+$ cell yield and recovery (using original controlled-rate cryopreservation), with ultimate goal of improving conditions for complete and long-term hematopoietic reconstitution after autologous SC transplants or therapy in the field of regenerative medicine (Table 2.1).

As mentioned, the most common hematological malignancies treated by SCs are different leukemias (nowadays mainly acute myeloid leukemia and Acute lymphoblastic leukemia), Hodgkin's disease or non-Hodgkin's lymphoma, multiple myeloma, and myelodysplastic or myeloproliferative disorders. In addition, a number of neoplasms and nonmalignant diseases were treated with SC transplants such as breast, ovarian, and testicular cancer, Wilm's tumor, neuroblastoma, as well as SAA, SCID, thalassemia, and various congenital or autoimmune disorders.

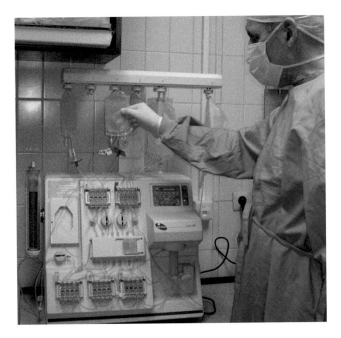

Fig. 2.6 Positive/negative immunomagnetic cell selection

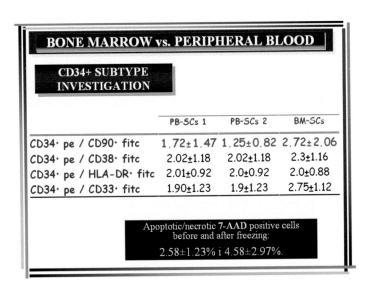

BONE MARROW vs. PERIPHERAL BLOOD			
CD34+ SUBTYPE INVESTIGATION			
	PB-SCs 1	PB-SCs 2	BM-SCs
CD34⁺ pe / CD90⁺ fitc	1.72±1.47	1.25±0.82	2.72±2.06
CD34⁺ pe / CD38⁺ fitc	2.02±1.18	2.02±1.18	2.3±1.16
CD34⁺ pe / HLA-DR⁺ fitc	2.01±0.92	2.0±0.92	2.0±0.88
CD34⁺ pe / CD33⁺ fitc	1.90±1.23	1.9±1.23	2.75±1.12

Apoptotic/necrotic **7-AAD** positive cells before and after freezing:
2.58±1.23% i 4.58±2.97%.

Table 2.1 Flow cytometry of SCs—CD34 subset quantity and ratio
PB-SCs 1 = chemotherapy + rHuG-CSF; PB-SCs 2 = rHuG-CSF alone; BM-SCs = BM derived cells

However, specific clinical aspects—such as optimal transplantation timing, therapeutic efficacy, complications—of the treatment of these disorders will not be discussed in this book.

Finally yet importantly, the clinical use of PB-SCs has a specific immune-mediated beneficial antitumor effect, particularly in hematooncology. Namely, it has long been known that the administration of the donor-specific lymphocytes (DSLs) resulted with an obvious immunomodulating effect. However, only recently have systems been developed to "separate" the Graft-versus-Leukemia (GvL) effect from GvHD—the best results were obtained in treatment of CML [1, 46]. Our investigations of the use of DSLs also confirmed efficacy of this treatment in patients with Philadelphia-positive CML relapsed after BMT. At the same time, our original in vitro test (named as "Test of Mixed Progenitors") was introduced to predict the clinical outcome of DSL treatment. These patients did not develop GvHD and currently they remain well in complete remission (direct evidence of the GvL effect) [47].

Umbilical Cord Blood Derived Stem Cell Usage

Patient's requests for SCs have only in ≤ 30 % (related) and ≤ 70 % (unrelated) possibility of finding an adult allogeneic donor. Because of the limited availability of donors, attention has turned to alternative sources of HLA-typed SCs. In recent years, UCB has emerged as a feasible alternative source of transplantable CD34$^+$ cells for allogeneic transplant, mainly in patients who lack HLA-matched donors of BM or PB derived SCs [1–3, 23–25]. UCB is relatively rich in "more primitive" SCs that can be used not only to reconstitute the hematopoietic system, but have the potential to give rise to non-hematopoietic cells (myocardial, neural, and endothelial cells, etc.) by transdifferentiation (Fig. 2.7).

The "naive" nature of UCB lymphocytes also permits the use of HLA-mismatched grafts at 1–2 loci without higher risk for severe GvHD relative to BM transplant from a full matched unrelated donor. On the other hand, UCB is rich in primitive NK cells, which possess impressive proliferative and cytotoxic capacities and can be induced Graft versus Leukemia (GvL) effect. The use of UCB is an accepted cell source for pediatric patients for whom smaller cell count is enough for engraftment, and for whom a matched unrelated allogeneic BM or peripheral blood SC donor is unavailable. However, a higher risk of graft failure was noticed in children weighing ≥ 45 kgmb. Since the number of SCs in UCB is limited and the collection can occur only in a single occasion—its use in adult patients can be more problematic [23–25].

UCB volume is typically 100 mL (range 40–240 mL) with a TNC count around 1×10^9 and CD34$^+$ approximately 3×10^6 per unit. UCB can easily be cryopreserved, thus allowing for the establishment of HLA-typed SC banks. Because UCB derived SC banking requires high financial investment and organizational efforts, banking efficiency should be optimized. An important determinant of banking efficiency is the ratio of collections that can be cryopreserved and supplied for transplant.

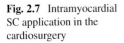
Fig. 2.7 Intramyocardial
SC application in the
cardiosurgery

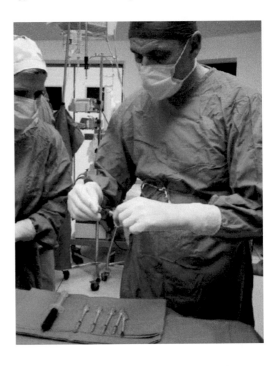

Although there were reasons for removing UCB units that may be less amenable to improvement, such as low volumes and low cell counts, a number of obstetric factors influencing the outcome of collections could be evaluated further, including the time of cord clamping, length of gestation, length of labor, the newborn's body weight, and the weight of the placenta [1–3, 24, 36].

Long-Term Storage of Stem Cells by Cryopreservation

Cryobiology is a scientific discipline that estimates the effects of ultra-low temperature on cell integrity and functionality in the "refrigerated biological system" and determines data applicable in cryopractice. Cryopreservation is beneficial when cells appear to be biologically, chemically, or thermally unstable after liquid-state storage. Its primary aim is to obtain both better cell recovery and postthaw viability. Thus, cryopreservation includes specific approaches/techniques designed to extend "therapeutic shelf life" of the cells (prolonged storage time) and to obtain minimum thermal damages (cryoinjury) [1–3, 26–30].

The use of cryobiology for living cell preservation began in 1949 with the freezing of bovine sperm cells, using glycerol as a cryoprotectant [48]. Afterwards, glycerol and dimethyl sulfoxide (DMSO) techniques were applied for cryopreservation of different blood-derived (progenitor or mature) cells [44, 49–56]. The basic goal of these initial cryoinvestigations was to predict the cell response to freeze/thaw processes and cryoprotectant addition/removal. However, evaluation of certain cryobiological variables (biophysical, physicochemical, and other events/parameters responsible for cryoinjury), as well as standardization of conceptual and practical aspects of cryopreservation, is still a question of considerable interest to researchers and practitioners [1–3, 27, 57–60].

As mentioned, SC cryopreservation is now in routine practice, but recent cryoinvestigations suggest that freezing strategies should be revised to optimize specific cryobiosystems, i.e., to minimize the cryoinjuries and maximize cell recovery. Microprocessor-restricted (controlled-rate) freezing is a time-consuming process, which requires high-level technical expertise. Uncontrolled-rate ("dump-freeze" without programmed cooling rate) technique is less costly because it does not require a programmed freezing device. However, the controlled-rate method is a high-class alternative to the uncontrolled-rate technique due to superior quantitative, morphological, ultrastructural, and functional cell recovery [28–30, 44].

Cryoinjuries can be detected as cell lesions, caused by the decrease of selected functions to the total cell destruction, i.e., cytolysis. At present it is considered that cryoinjuries result from the extensive volume reduction (cellular dehydration or solution effect) and/or massive intracellular ice crystallization (mechanical damage). Although independent, these mechanisms can also act together. The first event is expressed primarily at low-rate (≤ 10 °C/min) freezing, and the second one in high-rate (≥ 10 °C/min) freezing [2, 28, 56].

Therefore, determination of an optimal freezing rate (specific for each cell type and cryobiosystem) should be considered. It is the speed of cooling high enough to prevent cell dehydration and adequately low to make possible efflux of water from the cell. It would be ideal to find a cooling rate just less than the one, which causes intracellular crystallization [10, 27]. Optimal freezing rate is the function of the ratio between cell surface and volume, as well as of cell membrane permeability for water and its corresponding temperature coefficient—but it also depends on what type of cryopreservation strategy is applied. Last but not least, a higher degree of cell destruction has occurred when transition period from liquid to solid phase (fusion heat releasing) is prolonged (Fig. 2.8). The released heat of fusion— if not considered during controlled-rate freezing—could result in additional temperature fluctuation. That is why the period of transformation from liquid to solid phase will be prolonged, and its duration is directly related to the degree of cryoinjury [2, 26, 28].

Determination of the optimal freezing approach is essential, but it cannot solve all problems related to cell cryoinjury. To be precise, postthaw cell recovery and viability are high only when cryoprotectants are present in the cryobiosystem. They prevent or reduce the degree of cell thermal damages. In brief, cryoprotectants can express protective effect by the reduction of cell dehydration, as well as by decreasing the intensity of intracellular crystallization. However, they cannot protect the cell from an already existing excessive dehydration or from the effect of already formed intracellular ice crystals [26].

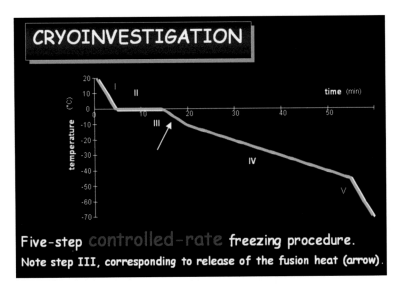

Fig. 2.8 Controlled-rate cryopreservation with compensation of the released fusion heat

Generally, cryoprotectants can be classified into nonpenetrating or extracellular and penetrating or intracellular compounds. Mechanisms of their action are complex and only partially recognized. Due to the differences in its chemical and other properties, it is not possible to discover a cryoprotective mechanism common for all cryoprotectants. In brief, extracellular agents could protect cells during high-rate freezing, reducing the intracellular ice crystal formation. On the other hand, intracellular cryoprotectants could provide protection in the course of low-rate freezing, decreasing the degree of cell dehydration [48–51, 61–64].

In practice, bone marrow SC cryopreservation consists of the following steps: (a) graft purification (if it is needed); (b) equilibration (cell exposure to cryoprotectant) and freezing; (c) cell storage at -90 ± 5 °C (mechanical freezer), at temperature from -120 to -150 °C (mechanical freezer or steam of nitrogen) or at -196 °C (liquid nitrogen); and (d) cell thawing in a water bath at 37 ± 3 °C. Cryopreservation of PB-SC has to be adapted to conditions which depend on the: (a) higher blood cell count; (b) presence of plasma proteins; (c) absence of lipid and bone particles in HSPC concentrate [2, 28].

Immediately after thawing, cells are transfused through a central vein catheter. Generally, patients tolerate the infusion of unprocessed SCs well, with no side effects. However, the grade of the potential reinfusion-related toxicity is associated with DMSO quantity in the cell concentrate [2]. Alternatively, cryoprotectant can be removed by washing, but this procedure results in substantial cell loss. The integrity of residual granulocytes is compromised within cryopreserved HSPCs and consequential DNA release during the thawing procedure may lead to cell "clumping" with resulting extra cell loss. To avoid this problem, a washing protocol by recombinant human deoxyribonuclease (rHu-DNase) is recommended [2,

65]. The addition of rHuDNase to cell concentrate seemingly proves to be effective in preventing "clumping" and it does not cause decreased expression of adhesion molecules, although it is not free of potential risks for patients. Moreover, the use of specific additives (e.g., membrane stabilizers) could improve postthaw cell recovery and it is probably a more effective approach than the decrease of DMSO concentration [28, 65].

Our results are in agreement with the abovementioned studies. Namely, we have found that the recovery of pluripotent and committed haematopoietic progenitors (CFU-Sd12 and CFU-GM) in the presence of 5 % vs. 10 % DMSO is superior [26]. However, it has also been demonstrated that the recovery of very primitive pluripotent haematopoietic stem cells (Marrow Repopulating Ability—MRA) is better when 10 % DMSO is used. These results imply a different "cryobiological request" of MRA cells in comparison with the nucleated cells and progenitors. Moreover, we have demonstrated that differences in cell recovery are not related to the changes in the total number of frozen/thawed cells, regardless of the use of cryopreservation strategy [26]. As a final point, our clinical studies showed that therapeutic use of the controlled-rate cryopreserved SCs in treatment of leukemias (ALL, ANLL, CML), multiple myeloma, Hodgkin's and non-Hodgkin's lymphoma, breast and ovarian cancer, and extragonadal non-seminal germ cell tumor resulted with high cell recovery (91 %) and rapid posttransplantation haematopoietic reconstitution (on the 11th day in average) [1, 9, 10, 44, 45].

*
* *

The intensification of myeloablative therapy with SC rescue, as well as increase in the use of allogeneic transplantations and different cell-mediated therapeutic approaches have resulted in higher needs for both SCs and practical operating procedures in minimizing cell damage during collection and cryostorage. SC collection systems have to effectively protect the biological and physical properties of cells that can be altered radically by the harvesting and/or purification processes, and have to include techniques and materials appropriate for human use. Although cryopreservation of SCs represents one complex process and the number of potential questions is higher than the number of possible answers, considering ever increasing needs for SCs as well as a rising use of different cell-mediated curative methods, it should also find its appropriate place in the current medical practice.

References

1. Pavlovic M, Balint B (2013) Stem cells and tissue engineering. Springer, New York
2. Balint B, Vucetic D, Ostojic G, Ljubenov M (2015) The basic concepts of transfusion medicine with hemobiology. MMA & Media Center Defense, Belgrade
3. Balint B, Stamatovic D, Andric Z (2004) Stem and progenitor cell transplantation. In: Balint B (ed) Transfusion medicine. CTCI, Belgrade, pp 525–547
4. Balint B, Radovic M, Lj M (1988) Bone marrow transplantation. Vojnosanit Pregl 45: 195–201

5. Thomas ED, Lochte HL (1959) Supralethal whole body irradiation and isologous marrow–transplantation in man. J Clin Invest 38:1709–1716
6. Mathe G, Jammet H (1959) Transfusions et greffes de moelle osseusse homologue chez des humaines irradies a haute danse accidentellement. Rev Fr Etud Clin Biol 4:226–238
7. Goldman JM, Th'ng KH, Park DS, Spiers ASD, Lowenthal RM, Ruutu T (1978) Collection, cryopreservation and subsequent viability of haemopoietic stem cells intended for treatment of chronic granulocytic leukemia in blast–cell transformation. Br J Haematol 40:185–195
8. Champ Itala M, Aho H, Remes K (2004) Reduced–intensity conditioning and blood stem cell transplantation from an HLA–identical sibling for severe aplastic anaemia: two patients with successful engraftment but a fatal post–transplant lymphoproliferative disorder in the other. Hematol J 5:440–443
9. Marjanovic S, Stamatovic D, Lj T, Tarabar O, Elez M, Balint B et al (2011) Autologous stem cell transplantation in the treatment of multiple myeloma—single center experience. Vojnosanit Pregl 68:387–392
10. Balint B, Stamatovic D, Todorovic M, Elez M, Vojvodic D, Pavlovic M, Cucuz–Jokic M (2011) Autologous transplant in aplastic anemia: quantity of CD34+/CD90+ subset as the predictor of clinical outcome. Transfus Apher Sci 45:137–141
11. Lazarus HM, Pecora AL, Shea TC, Koc ON, White JM, Gabriel DA et al (2000) CD34+ selection of hematopoietic blood cell collections and autotransplantation in lymphoma: overnight storage of cells at 4 degrees C does not affect outcome. Bone Marrow Transplant 25:559–566
12. Platzbecker U, Ehninger G, Bornhauser M (2004) Allogeneic transplantation of CD34+ selected hematopoietic cells—clinical problems and current challenges. Leuk Lymphoma 45:447–453
13. Saiz A, Graus F (2002) Transplantation of haematopoietic stem cells in multiple sclerosis. Rev Neurol 35:1136–1141
14. Smogorzewska EM, Weinberg KI, Kohn DB (2003) Transplantation of genetically modified cells in the treatment of children with SCID: great hopes and recent disappointments. Med Wieku Rozwoj 7:27–34
15. Gaziev J, Lucarelli G (2005) Stem cell transplantation and gene therapy for hemoglobinopathies. Curr Hematol Rep 4:126–131
16. Lasky LC, Ash RC, Kersey JH (1982) Collection of pluripotential hematopoietic cells by cytapheresis. Blood 59:822–827
17. Sato N, Sawada K, Takahashi TA, Mogi Y, Asano S, Koike T et al (1994) A time course study for optimal harvest of peripheral blood progenitor cells by granulocyte colony-stimulating factor in healthy volunteers. Exp Hematol 22:973–978
18. Balint B, Ivanovic Z (1995) Peripheral blood stem and progenitor cell transplantation. Bull Transfus 24:5–11
19. Moog R (2004) Apheresis techniques for collection of peripheral blood progenitor cells. Transfus Apher Sci 31:207–220
20. Martin I, Albert A, Alcorta I, Estella J, Rives S, Toll T et al (2003) Large volume leukapheresis for peripheral blood stem cell collection in children under 10 kg in weight. Bone Marrow Transplant 31:517–518
21. Moreiras–Plaza M, Albo C, Ares C (2004) Efficacy and safety of femoral vascular access for peripheral blood stem cell (PBSC) collection. Bone Marrow Transplant 33:347–350
22. Saif MW, Leitman SF, Cusack G, Horne M, Freifeld A, Venzon D et al (2004) Thromboembolism following removal of femoral venous apheresis catheters in patients with breast cancer. Ann Oncol 15:1366–1372
23. Rogers I, Casper RF (2004) Umbilical cord blood stem cells. Best Pract Res Clin Obstet Gynaecol 18:893–908
24. Skoric D, Balint B, Petakov M, Sindjic M, Rodic P (2007) Collection strategies and cryopreservation of umbilical cord blood. Transfus Med 17:107–113
25. Cohen Y, Nagler A (2004) Umbilical cord blood transplantation—how, when and for whom? Blood Rev 18:167–179

26. Balint B, Ivanovic Z, Petakov M, Taseski J, Jovcic G, Stojanovic N et al (1999) The cryo-preservation protocol optimal for progenitor recovery is not optimal for preservation of MRA. Bone Marrow Transplant 23:613–619

27. Balint B, Ljubenov M, Stamatović D, Todorović M, Pavlović M, Ostojić G et al (2008) Stem cell harvesting protocol research in autologous transplantation setting: large volume vs conventional cytapheresis. Vojnosanit Pregl 65:545–551

28. Balint B (2004) Coexistent cryopreservation strategies: microprocessor-restricted vs. uncontrolled-rate freezing of the "blood-derived" progenitors/cells. Blood Banking Transfus Med 2:62–71

29. Gorin NC (1986) Collection, manipulation and freezing of haemopoietic stem cells. Clin Haematol 15:19–48

30. Rowley SD (1994) Secondary processing, cryopreservation, and reinfusion of the collected product. In: Kessinger A, McMannis JD (eds) Practical considerations of apheresis in peripheral blood stem cell transplantation. Cobe BCT, Lakewood, pp 53–62

31. Tabilio A, Falzetti F, Zei T, De Ioanni M, Bonifacio E, Battelli F et al (2004) Graft engineering for allogeneic haploidentical stem cell transplantation. Blood Cells Mol Dis 33:274–280

32. Tabbara IA (1996) Allogeneic bone marrow transplantation: acute and late complications. Anticancer Res 16:1019–1026

33. Wilson C, Sylvanus T (2005) Graft failure following allogeneic blood and marrow transplant: evidence–based nursing case study review. Clin J Oncol Nurs 9:151–159

34. Kansu E (2004) The pathophysiology of chronic graft–versus–host disease. Int J Hematol 79:209–215

35. Yeh ETH, Zhang S, Wu HD, Kërbling M, Willerson JT, Estrov Z (2003) Transdifferentiation of human peripheral blood CD34+-enriched cell population into cardiomyocytes, endothelial cells, and smooth muscle cells in vivo. Circulation 108:2070–2073

36. Balint B (2004) Stem cells—unselected or selected, unfrozen or cryopreserved: marrow repopulation capacity and plasticity potential in experimental and clinical settings. Makedon Med Pregl 58(Suppl 63):22–24

37. Balint B, Obradovic S, Todorovic M, Pavlovic M, Mihaljevic B (2013) Stem cell-based (auto) grafting: from innovative research toward clinical use in regenerative medicine. In: Alimoghaddam K (ed) Stem cell biology in normal life and diseases. InTech, Rijeka, pp 111–135

38. Wagers AJ, Sherwood RI, Christensen JL, Weissman IL (2002) Little evidence for developmental plasticity of adult hematopoietic stem cells. Science 297:2256–2259

39. Beltrami AP, Urbanek K, Kajstura J, Yan S-M, Finato N, Bussani R et al (2001) Evidence that human cardiac myocites divide after myocardial infarction. N Engl J Med 344:1750–1757

40. Müller P, Pfeiffer P, Koglin J et al (2002) Cardiomyocites of noncardiac origin in myocardial biopsies of human transplanted hearts. Circulation 106:31–35

41. Shintani S, Murohara T, Ikeda H, Uenoi T, Honma T, Katoh A et al (2001) Mobilization of endothelial progenitor cells in patients with acute myocardial infarction. Circulation 103:2776–2779

42. Hassink RJ, de la Rivere AB, Mummery CL, Doevendans PA (2003) Transplantation of cells for cardiac repair. J Am Coll Cardiol 41:771–777

43. Rafii S, Lyden D (2003) Therapeutic stem and progenitor cell transplantation for organ vascularization and regeneration. Nat Med 9:702–712

44. Balint B, Todorovic M, Jocic M, Stamatovic D (2011) Haemotherapeutic approach to organ and cell transplants. ART, Belgrade

45. Balint B, Trkuljić M, Todorović M (2010) Basic principles of the haemotherapy. Cigoja, Belgrade

46. Kolb HJ, Schmid C, Chen X, Woiciechowski A, Roskrow M, Weber M et al (2003) Adoptive immunotherapy in chimeras with donor lymphocytes. Acta Haematol 110:110–120

47. Petakov M, Balint B, Bugarski D, Jovcic G, Stojanovic N, Vojvodic D et al (2000) Donor leukocyte infusion—the effect of mutual reactivity of donor's and recipient's peripheral blood

mononuclear cell on hematopoietic progenitor cells growth. Vojnosanit Pregl 57(5 Suppl):89–93

48. Polge C, Smith AU, Parkes AS (1949) Revival of spermatozoa after vitrification and dehydration at low temperatures. Nature 164:666

49. Mollison PL, Sloviter HA (1951) Successful transfusion of previously frozen human red cells. Lancet 2:862

50. Barnes DWH, Loutit JF (1955) The radiation recovery factor: preservation by the Polge–Smith–Parkers technique. J Natl Cancer Inst 15:901–906

51. Lovelock JE, Bishop MWH (1959) Prevention of freezing damage to living cells by dimethylsulfoxide. Nature 183:1394–1395

52. Meryman HT (1956) Mechanics of freezing in living cells and tissues. Science 124:515–521

53. Mazur P (1966) Theoretical and experimental effects of cooling and warming velocity on the survival of frozen and thawed cells. Cryobiology 2:181–192

54. Lewis JP, Passovoy M, Trobaugh FE (1966) The transplantation efficiency of marrow cooled to −100 °C at 2 °C per minute. Cryobiology 3:47–52

55. Litvan GG (1972) Mechanism of cryoinjury in biological systems. Cryobiology 9:181–191

56. Balint B, Radovic M (1993) Biophysical aspects of cryopreservation of hematopoietic stem and progenitor cells. Bull Transfus 21:3–8

57. Montanari M, Capelli D, Poloni A et al (2003) Long-term hematologic reconstitution after autologous peripheral blood progenitor cell transplantation: a comparison between controlled–rate freezing and uncontrolled–rate freezing at 80 degrees C. Transfusion 43:42–49

58. Lakota J, Fuchsberger P (1996) Autologous stem cell transplantation with stem cells preserved in the presence of 4.5 and 2.2% DMSO. Bone Marrow Transplant 18:262–263

59. Rowley SD, Feng Z, Chen L et al (2003) A randomized phase III clinical trial of autologous blood stem cell transplantation comparing cryopreservation using dimethylsulfoxide vs dimethylsulfoxide with hydroxyethylstarch. Bone Marrow Transplant 31:1043–1051

60. Balint B (2000) From the initial efforts of cell freezing to the standardization of the blood cell cryopreservation. Bull Transfus 46:3–8

61. Balint B, Todorovic M (2010) Stem cells—critical hemobiological aspects and clinical practice. Anest Reanim Transfus 38:9–22

62. Stiff PJ, Koester AR, Weidner MK et al (1987) Autologous bone marrow transplantation using unfractionated cells cryopreserved in dimethylsulfoxide and hydroxyethyl starch without controlled-rate freezing. Blood 70:974–978

63. Clark J, Pati A, McCarthy D (1991) Successful cryopreservation of human bone marrow does not require a controlled–rate freezer. Bone Marrow Transplant 7:121–125

64. Halle P, Tournilhac O, Knopinska-Posluszny W et al (2001) Uncontrolled–rate freezing and storage at −80 degrees C, with only 3.5–percent DMSO in cryoprotective solution for 109 autologous peripheral blood progenitor cell transplantations. Transfusion 41:667–673

65. Beck C, Nguyen XD, Kluter H, Eichler H (2003) Effect of recombinant human deoxyribonuclease on the expression of cell adhesion molecules of thawed and processed cord blood hematopoietic progenitors. Eur J Haematol 70:136–142

Chapter 3
A Concept of Cancer Stem Cells: Entity and Theories

If the facts don't fit the theory, change the facts.

Albert Einstein

Introduction

Cancer remains the second greatest cause of death in the United States with 574,689 casualties in 2012 according to the Centers for Disease Control and Prevention [1], leading much of medical funding to be allocated toward cancer research. Table 3.1 shows a breakdown in millions of dollars of research money allocated to the specific types of cancers, as well as the category in general (row 3) [2].

While many cancer researchers continue to build upon existing models of chemotherapy and radiation in efforts to improve their effectiveness by increasing their specificity, a separate cadre of researchers is exploring a more novel route: cancer stem cells (CSCs) [3–5].

What Are Cancer Stem Cells?

Healthy tissues come from a specific stem line originating from a pluripotent cell, much as red blood cells are made from hematopoietic stem cells (HSC). Crucial characteristics of these cells are the ability of self-renewal, differentiation through proliferation and plasticity (transdifferentiation capacity) [6–8]. All of these cells are described as having "stemness," a general term referring to two main characteristics: a state of highly promiscuous gene-expression patterns with particular phenotype designated by stem-cell markers, and the ability to transdifferentiate into specialized cell types (e.g., functionality) [9, 10]. This provides potential, observable outcomes, but no specialization (Fig. 3.1).

Now apply this same model but assume that something goes wrong along the differentiation pathway. Due to the genetic plasticity of these stem cells, a certain feedback pathway is augmented to no longer respond to environmental cues that would inhibit growth in the presence of other cells, or likewise, not respond to cues halting further division of a cell [10–12].

© The Author(s) 2015
M. Pavlovic, B. Balint, *Bioengineering and Cancer Stem Cell Concept*,
DOI 10.1007/978-3-319-25670-2_3

Table 3.1 Estimates of funding for Various Research, Condition, and Disease Categories (RCDC). ©2013 National Institute of Health

Research/disease areas (dollars in millions and rounded)	FY 2009 actual (non-ARRA)	FY 2009 actual (ARRA)10/	FY 2010 actual (non-ARRA)	FY 2010 actual (ARRA) 10/	FY 2011 actual	FY 2012 actual	FY 2013 (estimated)	FY 2014 estimated
Brain cancer ($)	234	42	274	36	280	281	283	284
Breast cancer ($)	722	111	763	61	715	800	805	808
Cancer ($)	5629	1120	5823	803	5448	5621	5649	5671
Cervical cancer ($)	84	15	93	8	119	112	113	113
Colorectal cancer ($)	281	48	291	26	313	302	303	305
HPV and/or cervical cancer vaccines ($)	25	2	25	2	24	26	26	26
Liver cancer ($)	94	12	102	10	74	73	74	74
Lung cancer ($)	178	36	201	22	221	233	234	235
Ovarian cancer ($)	102	13	122	10	138	147	148	149
Pancreatic cancer ($)	±	±	±	±	112	127	128	129
Prostate cancer ($)	310	47	331	31	284	257	258	259
Uterine cancer ($)	25	4	26	4	40	42	42	42

*The table above displays the annual support level for various research, conditions, and disease categories based on grants, contracts, and other funding mechanisms used across the National Institutes of Health (NIH [2])

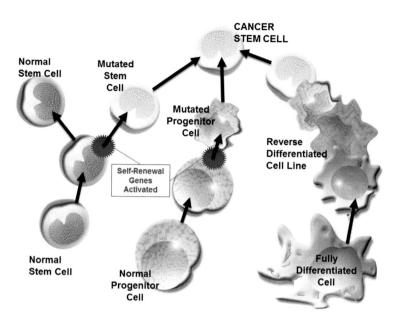

Fig. 3.1 Pathways of differentiation into cancer stem cells from different cell lines

Examples of differences and similarities in normal stem-cell populations and those of CSCs are shown in Table 3.2. The cells then grow and divide rapidly, symmetrically to give normal mature cells or differentiate by asymmetrical division into adult progeny cancer cells which initiate angiogenesis to provide their own blood (and inherently, nutrient) supply [12]. With the current treatment model, these "adult cells" that arose from the stem-cell line are the ones targeted for surgical resection or ablation, thus reducing the bulk of the neoplasm [13]. As the stem cells that created the tumor to begin with are so few in number, scans following treatment usually fail to identify populations of CSCs in this limited population. It further stands to reason that, following the original tumor's destruction, a second mass may evolve from the original stem line, and at times with a resistance to the original treatment (chemical or radiological) [11, 14–16].

There are two well defined yet different models of CSC within scientific community none of which completely can describe the features of CSC:

1. *Stochastic (clonal evolution) model*: This model states that all cancer cells hold tumorigenic potential and they are the product of clonal evolution by the acquisition of genetic mutations and epigenetic changes.
2. *Hierarchical model*: Tumors show hierarchy, with a subpopulation of cancer cells having a tumorigenic potential much greater than that of other cancer cells. Tumor contains hierarchical organization consisting of stem cells at top, which are cells within a tumor with the capacity to self-renewal and generate heterogeneous lineages of cancer cells, progenitors, and differentiated cells which are no longer able to produce tumors.

Table 3.2 Normal stem cells versus cancer stem cells

Endogenous & exogenous cues	Normal stem cells	Cancer stem cells	Reference(s)
Homeostatic regulation of tissue regeneration signals	Maintained	Lost	[4, 6, 8, 10]
Genetic plasticity	High	High	[3, 4, 8, 10, 12]
Growth regulation	Niche-driven	Self-sustained	[3, 4, 8, 14]
Antigrowth signal response	Yes	No	[6, 8]
Apoptosis signal response	Yes	No	[4, 6, 8, 15]
Limitation to replication	Yes	No	[4, 6, 8–10]
Angiogenic sustainability	No	Yes	[4, 6, 9]
Tissue invasion & metastasis	No	Yes	[4, 9, 10, 12]
Differentiation of resultant daughter cells	Yes	Impaired or none	[10, 18]
Aberrant DNA methylation	No	Yes	[12, 13, 21]
Anaerobic respiration	Yes	Yes	[4, 6, 10, 12, 17]
Heterogeneous population of cells	Yes	Yes	[4, 6, 7, 11, 13, 20]
Different sets of miRNA involved	Yes	Yes	[12, 13, 20, 26]

The new proposed model of cancer therapy that is explored in this paper is one where these CSCs are targeted and destroyed. Ideally, this would subsequently cause the tumor's mass to die given that its source for viability and regeneration would no longer exist. With as great of specificity as possible, this model promises little to no side-effects versus current treatment modalities [17–19].

Traditional stem cells and normal progenitor cells can have their self-renewal genes activated where functionality disappears in favor of replication as the cell's genome becomes more plastic. In addition, fully differentiated cells can devolve into a stem-cell line where self-differentiation becomes the cell's new niche.

Isolation and Characterization of Stem Cells

To provide the greatest level of specificity, thus the least amount of side-effects of a CSC-focused treatment, isolation, and specific characterization of these CSC populations must be done. In order to properly isolate CSCs from a heterogeneous population, they must be identified both in vivo and in vitro. In vivo—as a self-renewing population by xenotransplantation from humans into NOD/SCID mice, while in vitro, by their ability to proliferate in soft agar [20]. A problem exists as the current research shows that CSCs, same as normal stem cells are a heterogeneous population of cells (Table 3.2), are elusive and tend to be different in each instance of cancer. Because of this, CSCs must be determined by systems—biological tools that can identify patterns rather than through analysis of individual genes or even multiple gene-product behaviors, as this latter route is multivariable in its construct given the nature of stemness [20]. Stem cells from/of? a different origins/sources should be studied under conditions that are biased towards self-renewal as opposed to differentiation and should be analyzed at the nucleic-acid protein levels with a specific emphasis on posttranscriptional (DNA methylation) and posttranslational modifications (miRNA and proteins) and on changes in trafficking and localization of the molecules within the desired cells [20, 21]. Following this model, researchers have identified and catalogued several clonally expanded and perpetuated cancer stem-cell markers in efforts to identify, and subsequently target, these cells for destruction in in vivo models (Fig. 3.2).

The initial efforts of John Dick (1994) set the stage for an evidence-based concept of CSCs using the NOD/SCID mice model. He and his team successfully transplanted acute myelogenous leukemia (AML) stem cells from human patients into the murine model, wherein the stem cells regenerated AML in the mice [22].

In a study done by Li et al. (2007, 2010), pancreatic cancer cells were endowed in CD44+CD24+ESA+ cell population representing only 0.2–0.8 % of the total pancreatic cancer cell population. These cells, exhibiting the potential to initiate in vivo pancreatic cancer, are highly concentrated in this phenotypically defined population since as few as 100 CD44+CD24+ESA+ cells injected in vivo in immunosuppressed mice are forming tumors histologically indistinguishable from human

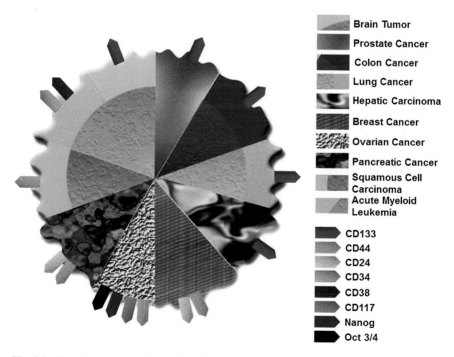

Fig. 3.2 Graphical representation of identified cancer stem-cell markers by cancer type (limited spectrum). *This graphic represents common stem-cell markers found by various researchers [21, 22, 24–26]

pancreatic cancer tumors from which they originated. The equivalent tumor potential could only be obtained with 100-fold more cells characterized by the CD44-CD24-ESA-phenotype (Fig. 3.3) [23, 24].

Similar features for human breast cancer cells were shown by Al-Hajj et al. In this case, the breast cancer cells were endowed by CD44+CD24-/low Lineage-phenotype. This cell population was revealed to be tumorigenic after in vivo injection to immunodeficient mice in the 100 cell dose, which is not the case for the alternate phenotypes even in doses of thousands of cells [25]. In similar preparation as Li et al. and their pancreatic cancer model, the 8-week-old NOD/SCID mice were used as the murine model for cancer cell implantation. Again, a xenograft of human breast cancer tissue was taken 1 h after surgical removal. The mincing, washing, and preparation was exactly as Li et al. had done so in the previous model (Fig. 3.4).

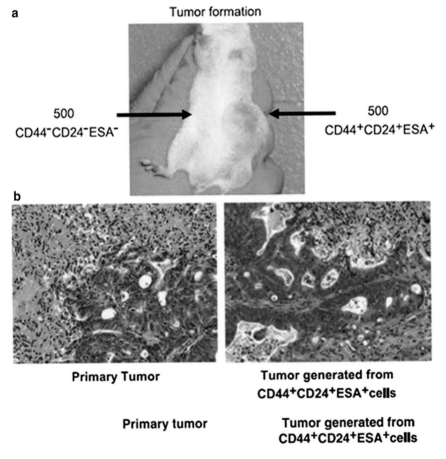

Fig. 3.3 Murine model outcome from Li et al. (**a**) Gross evidence of transplanted neoplasms by cell marker type. (**b**) Microscopic characterization of tumors by cell marker type

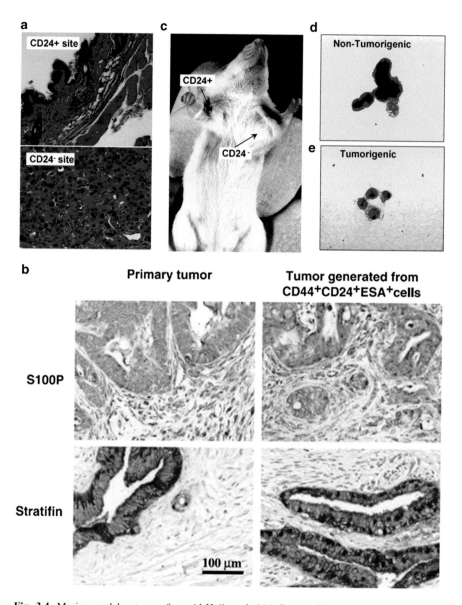

Fig. 3.4 Murine model outcome from Al-Hajj et al. *(**a**) Gross evidence of transplanted neoplasms by cell marker type. (**b**) Microscopic characterization of tumors by cell marker type

Since the beginning of 2013, further research has come forth wherein CSCs are targeted as the proposed cancer therapy in neoplasms such as breast cancer [26–29], myeloid leukemia [30], prostate cancers [31, 32], brain neoplasm [33] and colon cancer [34, 35]. The role of such therapy is analyzed in the colon CSC review by Puglisi et al. yet the results remained theoretical [35]. Another murine model study was found, similar to the ones undertaken by Li and Al-Hajj, where prostate CSCs were identified by their antigenic properties; however, in this study by Jachetti et al. the cell population was targeted and destroyed by cytotoxic T lymphocytes (CTLs) [32].

The next study that we found, however, was able to experiment in a human model, applying the previous murine model into a possible clinical application. In the Wilm's tumor (WT) model by Shukrun et al. a common renal pediatric solid tumor was formed from renal precursor cells that had failed to differentiate properly [36]. Shukrun and Shakked were able to identify the culprit progenitor cells (CSCs) via Neural Cell Adhesion Molecule 1 (NCAM1) and Aldehyde dehydrogenase 1 (ALDH1) enzymatic activity [12]. These WT NCAM1$^+$ cells were subsequently targeted using humanized NCAM1 antibody drug conjugate. Targeting of the human WT NCAM1$^+$ cell fraction with this conjugate resulted in the loss of WT CSCs, followed by complete tumor degradation.

In conclusion, the potential of targeting CSCs specifically in therapy models is proving to be promising. It is anticipated that this direction will be continued in conjunction with other current treatment modalities and will be manipulated to have greater cancer regression rates and better prognoses for the patients suffering these neoplasms.

Novel Treatment Modalities

The ultimate goal of cancer therapy lies in a few key ideas: (1) create as little side-effect of the treatment to the host's tissues; (2) treat as noninvasively as possible and (3) have long-term viability of treatment as stem cells vary in their genotypic expression. The third point may be moot if the CSCs can be obliterated in the first treatment, however.

As technological advancement creates new opportunity in other realms of science, so too does it in the world of medicine. Two new treatment modalities are on the forefront of oncological intervention: nanoparticle therapy and alternating magnetic fields on replicating CSCs. Although in the early phases of testing, the two show promise of accomplishing limited-to-no side-effects as well as being as non-invasive as possible. These novel treatments could be used in combination with chemotherapy or radiotherapy, as indicated.

A. The nanoparticle model has had difficulty from its inception with the use of quantum dots, later proven to be toxic. However, recent models have proven dramatically more effective as anticancer agents through various methods of delivery [37–42]. The model proposes the nanoparticles that are composed of silicon and can be coated with antibodies specific for stem-cell markers found on CSCs, or similar molecular conjugates. Through enhanced permeation retention, vascularized tumors would become infiltrated by these nanoparticles. Infrared radiation is then applied to the tumor, wherein the CSCs that now contained these particles would increase in heat at a much greater rate than cells not containing the particles, thus killing the targeted cells. Studies have been carried out for a variety of cancer types including rectal [37], liver cancer [39], prostate [40, 42], and breast [41] (Fig. 3.5).

B. The other novel therapy targeting CSCs that has shown great result is the use of alternating magnetic fields. As dividing cells undergo the various stages of cell replication, a developmental stage known as mitosis is the target of this therapy. During mitosis, all of the sister chromatids are lined up along the midline of the cell and still adjoined to one another by a centromere, which then become the target of spindles emergent from the centrioles at opposite poles of the dividing cell. These spindles have a polarity in charge due to their molecular composition. As this *Transient Treatment Field (TTF)*, is applied *via* an external array, the spindles of CSCs (which are smaller than other progenitors and normal cells) are disrupted by the alternating fields and a resultant disruption of cancer cell replication is accomplished [43, 44]. It is important to note, however, that CSCs are smaller than typical, normal-state mitotic cells and can therefore be targeted with specific frequencies in order to minimalize damage to healthy, non-cancerous cells (Fig. 3.6).

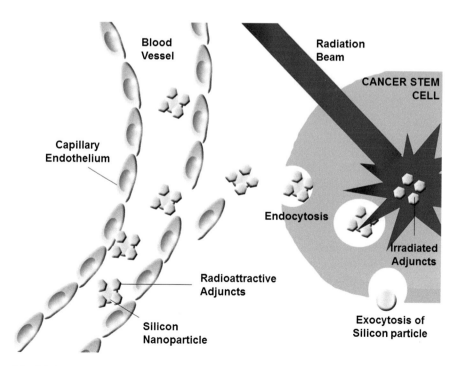

Fig. 3.5 Nanoparticle delivery of radioattractive adjuncts to cancer stem cells

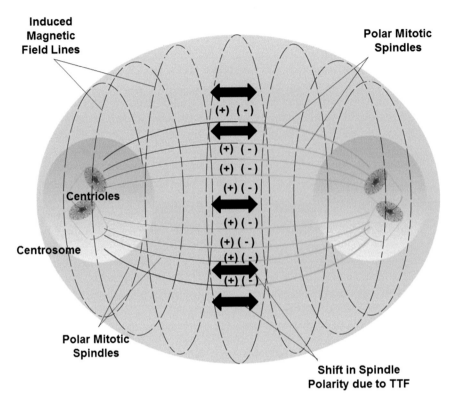

Fig. 3.6 Transient treatment field (TTF) inducing mitotic spindle rearrangement in polarity.
**The induced magnetic field specific to the frequency of CSCs disrupts the spindle formation and subsequent continuance of mitosis

The first clinical trial was in 2003 for patients with *glioblastoma (GBM)*, the most aggressive and most common form of primary brain tumor in the United States. Two years later, three of the original ten patients were still alive, two of which had no progression of the cancer whatsoever. In 2011, the FDA approved TTF as a viable treatment for GBM. Currently, clinical trials are being run for the utilization of TTF with lung cancer, as well as in vitro research for many other types of cancer, including cervical. Some devices already are produced and utilized in the market, making the treatment more readily available. NovocureTM, a commercial stage private oncology company, manufactures the device, NovoTTFTM-100A, a wearable device weighing around 6 lb that can fit into a shoulder bag for easy handling. Using noninvasive, insulated transducer arrays that are placed directly on the skin in the region surrounding the tumor, TTF therapy is unlike previous applications of electricity in medicine [44, 45].

Conclusions

Dynamic effects of therapeutic strategies directed against *CSCs* are obvious. It also seems possible, at least on the basis of the investigation of cancer stem-cell markers that certain populations of the cancer have for the cause the existence of CSCs, as some of them are for example, virally, chemically or radiation-induced. Cancer is today considered a spectrum of approximately 100 diseases with mostly not known causative factor. It is evident that a tumor tissue is a complex mix of cancer cells at various stages of differentiation, from uncommitted CSCs through different stages of cancer progenitor cells up to matured cancer cells, with a concomitant decrease in the levels of proliferative and/or metastatic potential. Both the CSC niche with supporting cell types and the matured cancer cell compartment create an intricate network of inter-dependency. Cancer therapy should ideally address both the CSCs and the matured cancer cells by slowing down proliferation and production of differentiated cancer cells and increasing apoptosis in both CSCs and matured cancer cells. In a fast-growing cancer, tumor therapy might come too late and/or be ineffective, or reduce tumor mass by killing matured cancer cells without targeting the CSC niche. The latter effect might stimulate CSC proliferation and increase the CSC pool, which would consequently result in a resurgence of even larger numbers of matured cancer cells. In another scenario, therapeutic intervention itself might provoke an enlargement of the CSC pool by selecting for more radio- and chemoresistant CSC clones. These CSCs will have a superior ability to repair DNA damage upon radiotherapy and/or overexpress members of the ABC transmembrane pumps, resulting in the swift efflux of certain chemotherapeutics. Over time, this new generation of CSCs could also include new mutant CSCs with even more aggressive signatures. CSC therapy targets the CSC niche itself by attenuating the self-replicating potential of CSCs and disturbing cellular crosstalk within the CSC niche. Increased apoptosis of CSCs will result in a significantly smaller number of matured cancer cells, which can then be addressed successfully with common anticancer therapies. Thus, anticancer therapy that only results in apoptosis of the matured cancer cells and/or only inhibits the proliferation of CSCs provides a potential window of opportunity for new and more aggressive CSC mutants to occur and might be unsuccessful if not dangerous. It is expected that the elimination of cancer caused by CSCs should target the CSC pool. So, successful treatment regimens would need to be the result of an orchestrated "target and destroy" effect, in those types of cancer where the reason for malignant transformation lies in existence of CSC within the particular tissue.

Through the identification of various research outcomes, it has become clear that two primary objectives lie ahead for the promulgation of the role of CSCs in future treatment modalities: (1) identification of protein surface markers/biomarkers on CSCs that uniquely identify them against normal cells in human tissue that thereby allow the development of therapeutic interventions that selectively target CSCs, and (2) characterization of CSCs in their ability to adapt to current forms of treatment so as to limit the proliferation of more resistant and dangerous forms of previous cancers through the identification of mutational "hot spots" in the aforementioned cancer stem-cell markers. While the promise of reduced cross-effect oncological treatment *via* CSC targeting gives great hope to the future of oncology and the patients that suffer from cancer, a great deal of work still remains. CSCs are a moving target and probably exist in such a small population that effective use of treatment modalities, although more promising than some miRNA studies and the like, still does not, in its current state, exist as a viable treatment option for all cancers. Much as microbiologists have difficulty in the world of fighting an ever-adapting organism, so to shall the oncologist and researchers that pursue this path. In combination with other therapies, however, it does appear that a reduction in risk associated with current treatment modalities would be evident. In light of the difficulty of the manipulation of the CSC model, the research that has been done thus far is providing a solid framework upon which a new, improved paradigm of oncological treatment will be established along with better understanding of this entity.

References

1. Centers for Disease Control and Prevention (CDC), and National Center for Health Statistics (2013) FastStats. Leading causes of death. Final 2010 data. CDC website
2. Estimates of funding for Various Research, Condition, and Disease Categories (RCDC), National Institute of Health (NIH) (2013) Web. http://report.nih.gov/categorical_spending.aspx. Accessed 6 Oct 2013
3. Jain M et al (2012) Highlights from recent cancer literature. Cancer Res 72:13
4. Nguyen LV et al (2012) Cancer stem cells: an evolving concept. Nat Rev Cancer 12(2):133–143
5. Al-Sarraf M, Martz K, Herskovic A et al (1997) Progress report of combined chemoradiotherapy versus radiotherapy alone in patients with esophageal cancer: an intergroup study. J Clin Oncol 15(1):277
6. Reya T, Morrison SJ, Clarke MF (2001) Stem cells, cancer, and cancer stem cells. Nature 414(6859):105
7. Delaney G, Jacob S, Featherstone C, Barton M (2005) The role of radiotherapy in cancer treatment. Cancer 104(6):1129
8. Ramalho-Santos M, Yoon S, Matsuzaki Y et al (2002) Stemness: transcriptional profiling of embryonic and adult stem cells. Science 298(5593):597
9. Zapori D (2004) The nature of stem cells: state rather than entity. Nat Rev Genet 5(11):873–878
10. Gil J, Stembalska A, Pesz KA, Sasiadek MM (2008) Cancer stem cells: the theory and perspectives in cancer therapy. J Appl Genet 49(2):193
11. Singh A, Settleman J (2010) EMT, cancer stem cells and drug resistance: an emerging axis of evil in the war on cancer. Oncogene 29(34):4741–4751
12. Lorico A (2011) Cancer stem cells. J Oncol 2011, 269437
13. Williams JL (2012) Cancer stem cells. Clin Lab Sci 25(1):50
14. Gupta PB et al (2009) Identification of selective inhibitors of cancer stem cells by high-throughput screening. Cell 138(4):645–659
15. Gupta PB, Chaffer CL, Weinberg RA (2009) Cancer stem cells: mirage or reality? Nat Med 15(9):1010–1012
16. Diehn M, Cho RW, Lobo NA et al (2009) Association of reactive oxygen species levels and radioresistance in cancer stem cells. Nature 458(7239):780–783

17. Rosen JM, Jordan CT (2009) The increasing complexity of the cancer stem cell paradigm. Science 324(5935):1670–1673
18. Shackleton M et al (2009) Heterogeneity in cancer: cancer stem cells versus clonal evolution. Cell 138(5):822–829
19. Hirsch HA et al (2009) Metformin selectively targets cancer stem cells, and acts together with chemotherapy to block tumor growth and prolong remission. Cancer Res 69(19):7507–7511
20. Jamieson CHM et al (2009) Methods of identifying and isolating stem cells and cancer stem cells. US Patent No. 7,622,255, 24 Nov 2009
21. Liu C et al (2011) The microRNA miR-34a inhibits prostate cancer stem cells and metastasis by directly repressing CD44. Nat Med 17(2):211–215
22. Lapidot T et al (1994) A cell initiating human acute myeloid leukaemia after transplantation into SCID mice. Nature 367:645–648
23. Li C, Heidt DG, Dalerba P et al (2007) Identification of pancreatic cancer stem cells. Cancer Res 67(3):1030
24. Li C, Lee CJ, Simeone DM (2009) Identification of human pancreatic cancer stem cells. In: Yu JS (ed) Cancer stem cells. Humana, New York, pp 161–173
25. Al-Hajj M, Wicha MS, Benito-Hernandez A, Morrison SJ, Clarke MF (2003) Prospective identification of tumorigenic breast cancer cells. Proc Natl Acad Sci U S A 100(7):3983
26. Shimono Y et al (2009) Downregulation of miRNA-200c links breast cancer stem cells with normal stem cells. Cell 138(3):592–603
27. Charafe-Jauffret E et al (2009) Breast cancer cell lines contain functional cancer stem cells with metastatic capacity and a distinct molecular signature. Cancer Res 69(4):1302–1313
28. Ginestier C et al (2010) CXCR1 blockade selectively targets human breast cancer stem cells *in vitro* and in xenografts. J Clin Invest 120(2):485
29. Neve RM, Chin K, Fridlyand J et al (2006) A collection of breast cancer cell lines for the study of functionally distinct cancer subtypes. Cancer Cell 10(6):515
30. Zhao C et al (2009) Hedgehog signalling is essential for maintenance of cancer stem cells in myeloid leukaemia. Nature 458(7239):776–779
31. Collins AT, Berry PA, Hyde C et al (2005) Prospective identification of tumorigenic prostate cancer stem cells. Cancer Res 65(23):10946
32. Jachetti E, Mazzoleni S, Grioni M et al (2013) Prostate cancer stem cells are targets of both innate and adaptive immunity and elicit tumor-specific immune responses. Oncoimmunology 2(5), e24520
33. Singh SK, Hawkins C, Clarke ID et al (2003) Identification of a cancer stem cell in human brain tumors. Cancer Res 63(18):5821
34. Vermeulen L et al (2010) Wnt activity defines colon cancer stem cells and is regulated by the microenvironment. Nat Cell Biol 12(5):468–476
35. Puglisi MA, Tesori V, Lattanzi W, Gasbarrini GB, Gasbarrini A (2013) Colon cancer stem cells: controversies and perspectives. World J Gastroenterol 19(20):2997
36. Shukrun R, Shakked NP, Dekel B (2013) Targeted therapy aimed at cancer stem cells: Wilm's tumor as an example. Pediatr Nephrol 29(5):815–823
37. Fisher B, Wolmark N, Rockette H, Redmond C et al (1988) Postoperative adjuvant chemotherapy or radiation therapy for rectal cancer: results from NSABP protocol R-011. J Natl Cancer Inst 80(1):21
38. Brannon-Peppas L, Blanchette JO (2004) Nanoparticle and targeted systems for cancer therapy. Adv Drug Deliv Rev 56(11):1649
39. Davis ME, Chen Z, Shin DM (2008) Nanoparticle therapeutics: an emerging treatment modality for cancer. Nat Rev Drug Discov 7(9):771
40. Gaitanis A, Staal S (2010) Liposomal doxorubicin and nab-paclitaxel: nanoparticle cancer chemotherapy in current clinical use. Methods Mol Biol 624:385–392
41. Gradishar WJ, Rjulandin S, Davidson N et al (2005) Phase III trial of nanoparticle albumin-bound paclitaxel compared with polyethylated castor oil-based paclitaxel in women with breast cancer. J Clin Oncol 23(31):7794

42. Farokhzad OC, Jon S, Khademhosseini A et al (2004) Nanoparticle-aptamer bioconjugates a new approach for targeting prostate cancer cells. Cancer Res 64(21):7668

43. Kirson ED, Gurvich Z, Schneiderman R et al (2004) Disruption of cancer cell replication by alternating electric fields. Cancer Res 64(9):3288

44. Stupp R, Wong ET, Kanner AA et al (2012) NovoTTF-100A versus physician's choice chemotherapy in recurrent glioblastoma: a randomised phase III trial of a novel treatment modality. Eur J Cancer 48:2192–2202

45. Dylla SJ, Park I-K, Gurney AL (2009) Cancer stem cells. In: Uma L, Chesnut JD, Bhaskar T (eds) Emerging technology platforms for stem cells. Wiley, Hoboken, p 129

Chapter 4
Cancer Stem Cell Markers: Classification and Their Significance in Cancer Stem Cells

A thinker sees his own actions as experiments and questions— as attempts to find out something. Success and failure are for him answers above all.

Friedrich Nietzsche

Introduction

Cancer is a widely spread disease in modern civilization or rather a spectrum of diseases with similar manifestations and outcome through the course of disease. The cause is usually unknown although in some forms of cancers it is defined: chemicals (bladder cancer), radiation (Leukemias), and viruses (Epstein–Bar virus/ Lymphoma, varicella virus/cervix uteri) [1–3]. In 2014, there were 1,665,540 new cancer cases diagnosed and 585,720 cancer deaths in the Unites States according to the American Cancer Society [1, 2]. Cancer remains the second most common cause of death in the United States, accounting for nearly 1 of every 4 deaths. Although researchers have continued to improve the level of effectiveness of standard oncology treatments, such as chemotherapy, radiotherapy, and surgical resection, they can only minimize the bulk tumor, which again, results in the tumor relapsing [3–9]. The theory of Cancer Stem Cells (CSC's) is gaining increasing attention from scientists and has, therefore, become an important focus of cancer research [10–19]. Understanding the properties of these CSCs is fundamental to future studies in this direction. Therapeutic strategies that focus on targeting CSCs labeled with specific/unique markers are expected to help address the ineffectiveness of traditional cancer therapies, which would otherwise result in therapy resistance and relapse of the disease. While a relatively vast amount of research has been done, there are still many controversies as well as ambiguity in regards to the CSC concept [19–25]. Through analysis of both past and present reports, we attempt to assess and present the current understanding of the CSC concept. By combining the use of conventional therapies with targeted CSC therapy, the cancer as a whole might be targeted and, as a result, offer a promising approach for a long-lasting treatment.

© The Author(s) 2015
M. Pavlovic, B. Balint, *Bioengineering and Cancer Stem Cell Concept*,
DOI 10.1007/978-3-319-25670-2_4

Background and Significance

CSCs are defined as a group of cells within a tumor that have the ability to self-renew, drive tumorigenesis, and are multipotent [12]. The hypothesis states that, although CSCs only represent a rare and subtle cell population within the tumor, their high tumorigenic capacity drives tumorigenesis [18]. CSCs proliferation produces more CSCs and in addition creates the differentiated cell types that constitute the bulk of the tumor due to their fundamental stem cell-like properties. Due to CSCs increased resistance to toxic and chemical substances, this subpopulation of cells is thought to be behind the cells' resistance to chemotherapy and disease relapse [20]. Cancer is a group of heterogeneous cells with respect to the level of tumorigenicity. However, unlike other cancer cells in a tumor, CSCs are capable of establishing new tumors when xenotransplanted into nonobese diabetic/severe immunodeficient animal models (NOD/SCID) [26–29]. Although it has been shown that cancer cells other than CSCs are able to proliferate at a faster rate than CSCs, they have slight tumor-initiating potential [12]. Therefore, it has been suggested that, to eradicate or elimi-nate a tumor or prevent it from reoccurring, all CSCs in that tumor must be eliminated.

There are two main models that describe cancer origin as well as metastatic progression.

1. *Stochastic (clonal evolution) model*: This model states that all cancer cells hold tumorigenic potential through accumulation of epigenetic changes and genetic mutations [3, 12].
2. *Hierarchical (CSCs) model*: Tumors represent a level of hierarchy organiza-tion, with a subpopulation of cancer cells known as CSCs holding a tumori-genic potential much greater than that of other cells contributing to the tumor [3, 12]. These CSCs reside from the same clone within the tumor and perpetu-ate same Cancer Stem Marker molecules during proliferation.

The existence of CSCs was first demonstrated in acute myeloid leukemia (AML) when John Dick was able to successfully complete a xenotransplantation of acute xylogenous leukemia stem cells (CD34$^+$ CD38$^-$) from human patients into NOD/SCID mice models [3]. AML is a heterogeneous disease in which a variety of distinct genetic abnormalities have been defined both clinically and biologically [3]. Despite the heterogeneity of cancer cells it was seen that these cells were able to regenerate acute myelogenous leukemia in those mice showing clonal origin and tumorigenicity. It was also seen that these cells were able to give rise to AML in secondary receivers indicating self-renewal capability (Table 4.1).

Furthermore, Al-Hajj et al. (2003) showed analogous results after xenotransplan-tation of human breast CSCs where injected into immunodeficient mice [30–33]. Six months after injection, the injection sites were examined and it was noted that the CD44+CD24−/lowLineage− injection sites contained tumors whereas the CD44+CD24+Lineage− injection sites contained no detectable tumors. The cell population, CD44+CD24−/lowLineage−, was discovered to be tumorigenic after in vivo injection to immunodeficient mice. In both cases only the CD44+CD24−/lowLineage− cells had the capability to proliferate and as a result form new tumors.

In an additional experiment conducted by Li et al. (2007) the cell population, CD44+CD24+ESA+, demonstrated the potential to initiate in vivo pancreatic cancer in NOD/SCID mice [26, 27]. This phenotypically defined population was perceived to be highly concentrated since as few as 100 CD44+CD24+ESA+ cells injected in nonobese severely immunosuppressed mice formed tumors histologically indistinguishable from human pancreatic cancer tumors from which they originated (Table 4.1).

Table 4.1 In vivo and in vitro experiments which have confirmed existence of unique cancer stem cells of different tissues (blood, breast, and pancreas)

Model	Experiment	Results
In vivo (NOD/SCID model)	Bonnet & Dick (1994); AML	Acute myelogenous leukemia stem cells, CD34+ CD38− were successfully transplanted from human patients into this immunodeficient model, in which the stem cells regenerated acute myelogenous leukemia in those mice
	Al-Hajj et al. Breast Cancer cells (2003);	Six months after injection, the injection sites were examined by histology. The CD44+CD24−/lowLineage− injection sites contained tumors whereas the CD44+CD24+Lineage− injection sites contained no detectable tumors. The cell population, CD44+CD24−/lowLineage−, was discovered to be tumorigenic after in vivo injection to immunodeficient mice. In both cases only the CD44+CD24−/lowLineage− cells had the capacity to proliferate to form new tumors
	Li et al. (2007); Pancreatic cancer cells	The cell population, CD44+CD24+ESA+, exhibits the potential to initiate in vivo pancreatic cancer and is highly concentrated in this phenotypically defined population since as few as 100 CD44+CD24+ESA+ cells injected in vivo in immunosuppressed mice are forming tumors histologically indistinguishable from human pancreatic cancer tumors from which they originated
In vitro		Cancer stem cells are identified within in vitro experiments via their ability to proliferate in agar. For example, propagation of breast cancer from three breast cancer lesions was completed with the derived culture of CD44+/CD24− cell. The cell population was able to give rise to new tumors of the same clone

Cancer Stem Cell Markers: Classification

Finding cancers' stem cells is a rapidly growing area of research [5, 7–11]. These cancer-causing cells, which make up a small fraction of cells within tumors, have properties similar, but also different from those of normal stem cells (Table 4.2) [5] CSCs make up only 1–2 % of the total cancer cells in a leukemia patient, which makes it challenging to find these CSCs from other cancer cells. Therefore, it seems that promising research can only be established, by further studying adult stem cells as well as embryonic stem cells (ESC). The ESC research still faces ethical problems and therefore is substantially controversial because an early embryo is destroyed when researchers remove stem cells from it. An alternative is to take the stem cells from embryos that carry a genetic defect for specific diseases.

Are Cancer Cells Transformed Normal Stem Cells?

Researchers have traditionally thought of cancer as a collection of cells, all growing exponentially. According to the new research, conventional cancer therapies do an effective job killing the majority of cancer cells within the tumor, but they may miss CSCs.

In Table 4.2 we have presented similarities and differences between normal and CSCs. The research in that direction is still going on. As a result, cancers often

Table 4.2 Comparison of normal stem cells versus cancer stem cells

Endogenous & exogenous cues	Normal stem cells	Cancer stem cells	Reference(s)
Homeostatic regulation of tissue regeneration signals	Maintained	Lost	[4, 6, 8, 10]
Genetic plasticity	High	High	[3, 4, 8, 10, 12]
Growth regulation	Niche-driven	Self-sustained	[3, 4, 8, 15]
Antigrowth signal response	Yes	No	[6, 8]
Apoptosis signal response	Yes	No	[4, 6, 8, 16]
Limitation to replication	Yes	No	[4, 6, 8–10]
Angiogenic sustainability	No	Yes	[4, 6, 9]
Tissue invasion & metastasis	No	Yes	[4, 9, 10, 12]
Differentiation of resultant daughter cells	Yes	Impaired or None	[10, 20]
Aberrant DNA methylation	No	Yes	[12, 13, 23]
Anaerobic respiration	Yes	Yes	[4, 6, 10, 12, 19]
Heterogeneous population of cells	Yes	Yes	[4, 6, 7, 11, 13, 22]
Different sets of miRNA involved	Yes	Yes	[12, 13, 22, 28]

DNA deoxyribonucleic acid, *miRNA* micro ribonucleic acid

reoccur. Even hematologic and some non-hematologic malignancies treated by autologous stem cell transplant and high dose chemotherapy have shown that regardless of survival rate of some cancers, the final outcome is death, due to recurrence of cancer.

As a result, cancers often reoccur. Even hematologic and some non-hematologic malignancies treated by autologous stem cell transplant and high-dose chemotherapy have shown that regardless of survival rate of some cancers, the final outcome is death, due to recurrence of cancer. The reason is (among others) in the fact that clinicians are injecting also cancer cells with healthy stem cells during re-infusion after apheresis collection, which accumulate and renew with a time to the critical level causing relapse or death [33, 34].

Table 4.3 presents (to our knowledge) the most completed/detailed classification of discovered CSC markers for different tissues and/or organs.

The teratocarcinomas are able to differentiate into normal mature cells when transplanted into another animal. This alternation between developmental and tumor cells status demonstrates how closely development and cancer are related. The present-day challenge is to decode the common molecular mechanism and genes involved in self-renewal for cancer cells and stem cells [34–39].

Cancer	CD24	CD44	ALDH1	CD90	CD29	CD117	CD133	Integrin	Integrin	CD166	Nanog	ABCG2	CD96	CD34	NCAM1	CD271	CD105	POL5F1	CD38	Reference(s)
Breast		■																		Hajj (2003); Ferro de Beca (2012)
Prostate							■	■												Maeda(2009); Hoogland (2013)
Colon	■						■													Vermeulen (2008)
Brain							■													Zeppernick (2008)
Lung							■													Bertollini (2009)
Pancreatic	■																			Li (2007), Zhu (2012)
Hepatic		■		■									■							Zhang (2013); Yan (2008); Shengyong (2007)
Ovarian						■	■													Luo (2011); Zhang (2011); Siu (2013)
Acute Myeloid Leukemia													■	■						Hosen (2007); Horton (2012); Bonnet (1997)
Wilm's Tumor		■													■					Shukrun (2013); Pode-Shakked (2012)
Melanoma		■														■				Luo (2012); Civenni (2011)
Gastric	■	■																		Takaishi (2009); Zhang (2011)
Renal																	■			Sandlund (2006; Bussolati (2008); Azzi (2011)
Thyroid																		■		Soon-Hyun (2013)

Cancer Stem Cell Markers

Table 4.3 Cancer stem cell markers according to different sources

Cancer Stem Cell Markers and Their Function in Normal and Cancer Stem Cells

CSCs markers have been gradually discovered and belong to proteins group of molecules known as cluster destination antigens (CDs). A very small number belongs to other categories such as: ALDH1, integrins ($\alpha2\beta$, $\alpha6$), Nanog, ABCGT, NCAM1, and POU5F1 [40–50]. Their function in normal and cancerous cell is different [40–47]. Table 4.3 represents our effort to give the newest classification of these markers according to their functional role in both normal and cancerous tissues as well as targeted therapeutic modality developed up to date [51–64].

There is a lot of work to be done on clarification of the eventual changes in these molecules in cancerous cells (and distinction with respect to CSCs alone) and causes that are inducing these changes. Apparently, normal CSCs are different as well as similar (Table 4.4). Does the fact that they have similarities can be "a smoking gun" for turning malignant into normal stem cells? We still do not know.

However, accumulation of the knowledge itself represents a solid basis for future research. Our goal was not to consider nor hypothesize the possibilities, but to collect and narrow down the known Cancer Stem Cell markers, e.g., to facilitate further research by offering the classified relevant markers and pointing out those aspects of relevant research that did not give the answers to fundamental questions such as:

- Why are these molecules playing different roles in cancerous tissues?
- Is their role different in particular CSCs?
- What is/are the causative factor(s) of these differences at a molecular level (conformation, activation, suppression of basic function, etc.)?
- How can we design targeted therapy based upon these answers?

Conclusions

The concept of CSCs is based upon Virchow's vision, and is not as new as we would think at first sight. The specific molecular markers (proteins) for CSCs in various cancers are identified and CSCs are tested in vitro and in vivo models/conditions. The role of these proteins in normal and cancerous cells differs. The causative factors of these changes at molecular level are unknown. The concept of targeted tumor therapy is developing, but requires more research in order to completely function.

Table 4.4 Cancer stem cell markers and their function in noncancer and cancer cells

Marker	Function in noncancerous tissue	Function in cancerous tissue	Targeted therapy modality	References
CD24	Modulates B-cell activation responses	Indirectly stimulates cell adhesion to fibronectin, collagens I and IV, and laminin through the activation of $\alpha3\beta1$ and $\alpha4\beta1$ integrin activity. Moreover, expression supported rapid cell spreading and strongly induced cell motility and invasion as well as induced proliferation and motility	Anti-CD24 monoclonal Ab (SWA11-ZZPE38 immunotoxin) decreases CD24 expression and inhibits cell proliferation	[51]
CD44	A cell-surface glycoprotein involved in cell–cell interactions, cell adhesion, and migration	Adhesion with hyaluronic acid plays an important role in cell migration, tumor growth, and progression	Monoclonal antibody called RG7356 specifically targeted CD44 and was directly toxic to cancer cells, but had little effect on normal B cells. H90 anti-CD44 therapy eradicated AML CSCs showing that it interfered with transport to a stem cell supportive microenvironment	[52, 53]
ALDH1	Catalyzes the oxidation of aldehydes	Higher expression results in higher rates of cell proliferation, microsphere formation, and migration	Adoptive therapy with ALDH1A1-specific CD8p T cells eliminated ALDH cells, inhibited tumor growth and metastases, or prolonged survival of xenograft-bearing immunodeficient mice	[56]
CD90	Plays a role in cell–cell or cell–ligand interactions during synaptogenesis and other events in the brain	Stimulates stronger proliferation and self-renewal abilities, and higher tumorigenic capacity	5-FU inhibits proliferative CD90+ cancer cells, decreases tumor size, and improves survival	[64]
CD29	Involved in cell adhesion and recognition	Metastatic diffusion of tumor cells and development		
CD117	Type 3 transmembrane receptor for mast cell growth factor and plays an essential role in the regulation of cell survival and proliferation	A proto-oncogene, meaning that overexpression or mutations of this protein can lead to cancer	Imatinib was confirmed to selectively target CD117(+) cells in vitro, and synergistically enhanced the antitumor effect of low-dose cisplatin in vivo	[55]

(continued)

Table 4.4 (continued)

Marker	Function in noncancerous tissue	Function in cancerous tissue	Targeted therapy modality	References
CD133	Maintains stem cell properties by suppressing differentiation, binds cholesterol in cholesterol-containing plasma membrane microdomains	Initiate tumor growth, produce tumors with preserved self-renewal and differentiation capabilities and without phenotypic alterations after serial transplantation	The viruses, termed MV-141.7 and MV-AC133, infected and selectively lysed CD133 tumor cells. Both viruses exerted strong antitumoral effects on human hepatocellular carcinoma growing in nonobese diabetic/severe combined immunodeficient (NOD/SCID) mice	[54]
α2β Integrin	Major collagen-binding integrin	Plays a role in melanoma cell migration on type IV collagen and laminin. In several studies, the expression of α2β was increased in metastatic cells when compared to cells in the primary tumor	Inhibition of the α2β1 integrin with function-blocking monoclonal antibodies inhibited tumor growth in SCID mice	[57]
α6 Integrin	Receptor for laminin on platelets	Changes in the pattern during malignant transformation are highly dependent on the type of the cancer. An altered pattern allows the cancer cells to recognize variable matrices, but it may also lead to altered signaling and changes in gene expression, increased expression of α6 integrin may contribute to the invasive capacity of the progression in cancer	A lentivirus delivering ShRNA directed against integrin α6 was transplanted into the brains of immunocompromised mice. Mice bearing integrin α6 shRNA showed significantly reduced tumor formation and greater median survival	[58]
CD166	Binds to T-cell differentiation antigen CD6, and is implicated in the processes of cell adhesion and migration	Protein expression is reported to be upregulated in a cell line deriving from metastasizing carcinoma, restricted to subsets of cells involved in dynamic growth and/or migration as well as tumor progression		

Nanog	A transcription factor critically involved with self-renewal of undifferentiated embryonic stem cells	Dysregulated and intimately involved in promoting tumorigenesis in part through regulation of the cancer stem cell (CSC) population, promotes the migration and invasion of breast cancer cells, implicated to promote chemoresistance	Delivery of RNAi-mediated Nanog knockdown was observed to reduce growth of subcutaneous xenograft tumors in vivo. They pre-transfected shNanog in cancer cell lines before injecting the cells in NOD/SKID mice and found that the shNanog-transfected cells presented reduced long-term clonogenic growth, proliferation, and tumor size in mice. [59]
ABCG2	Transport various molecules across extra- and intracellular membranes	Responsible for tumor self-renewal, overexpression confers high levels of resistance, preventing toxins from entering cells as well as potentially playing a role in regulating stem cell differentiation	The blocking of ABCG2 function by fumitremorgin C, a chemical inhibitor, inhibited cancer cell proliferation [60]
CD96	Plays a role in the adhesive interactions of activated T and NK cells during the late phase of the immune response. It may also function in antigen presentation	Stimulates cytotoxicity of activated NK cells, mutations result in the loss of adhesion and growth activities	
CD34	Possible adhesion molecule with a role in early hematopoiesis by mediating the attachment of stem cells to the bone marrow of extracellular matrix. Acts as a scaffold for the attachment of lineage-specific glycans	Promotes the proliferation and blocks the differentiation of progenitor cells, mutations have been associated with increased cancer aggressiveness	Naturally occurring parthenolide and synthesized demthylaminoparthenolide inhibitors induce apoptosis of CD34+. These results were notable because the in vitro observations were verified in vivo by treating mice previously engrafted with AML with DMAPT [55]

(continued)

Table 4.4 (continued)

Marker	Function in noncancerous tissue	Function in cancerous tissue	Targeted therapy modality	References
NCAM1	The encoded protein is involved in cell-to-cell interactions as well as cell matrix interactions during development and differentiation	This pro-malignant function is mediated by its interaction with fibroblast growth factor receptor; promotes the progression of cancer via FGFR signaling		
CD271	Plays a role in the regulation of the translocation of GLUT4 to the cell surface in adipocytes and skeletal muscle cells in response to insulin	Enhanced tumor-initiating capability, enable tissue invasion by breaking down the extracellular matrix		
CD105	Major glycoprotein of vascular endothelium. Involved in the regulation of angiogenesis and in addition plays a critical role in the binding of endothelial cells to integrins and/or other RGD receptors	A component of the receptor complex of Transforming Growth Factor (TGF)-β, a pleiotropic cytokine involved in cellular proliferation, differentiation, and migration	The anti-CD105 mab antibody conjugated with endoglin immunotoxin efficiently suppressed tumor growth in murine models bearing breast and colon carcinoma without any significant systemic side effects	[53]
POU5F1	Controls the expression of a number of genes involved in embryonic development	Promotes invasion and metastasis of some solid tumors through enhanced degradation of surrounding extracellular matrix	Lopinavir-induced apoptosis efficiently in Oct-4-positive CSCs and in addition reduced proliferation, clonogenicity, and selectively induced cell death in the CSC population, thus restraining CSC-induced allograft formation	[62]

CD38	Novel multifunctional ectoenzyme, functions in cell adhesion, signal transduction, and calcium signaling	A component of a molecular network delivering growth and survival signals to chronic lymphocytic leukemia cells, elevated expression is associated with advanced disease stage, high-risk cytogenetics, shorter lymphocytes doubling time (LDT), shorter time to initiation of first treatment (TFT), and poorer response to therapy	RA-induced CD38 antigen was found to target anti-CD38-based immunotoxin and effectively induced selective killing of leukemia cells	[61]

References

1. National Center for Health Statistics. Leading causes of death. http://www.cdc.gov/nchs/fastats/lcod.htm. Accessed 09 Oct 2013
2. National Center for Health Statistics. Estimates of funding for Various Research, Condition, and Disease Categories (RCDC). http://report.nih.gov/categorical_spending.aspx. Accessed 09 Oct 2013
3. Mayfield J, Pavlovic M (2014) A concept of cancer stem cells: the current understanding and a look ahead. Anest Reanim Transfuziol 41(1–2):27–31
4. Pavlovic M (2014) Very small embryonic like cells (VSELs): pros and cons—review and perspectives in the light of critical data and controversies. Anest Reanim Transfuziol 41(1–2):33–48
5. Jain M, Nilsson R, Sharma S, Madhusudhan N, Kitami T, Souza AL et al (2012) Highlights from recent cancer literature. Cancer Res 72:13
6. Nguyen LV, Vanner R, Dirks P, Eaves CJ (2012) Cancer stem cells: an evolving concept. Nat Rev Cancer 12(2):133–143
7. Al-Sarraf M, Martz K, Herskovic A, Leichman L, Brindle JS, Vaitkevicius VK et al (1997) Progress report of combined chemoradiotherapy versus radiotherapy alone in patients with esophageal cancer: an intergroup study. J Clin Oncol 15(1):277–284
8. Reya T, Morrison SJ, Clarke MF, Weissman IL (2001) Stem cells, cancer, and cancer stem cells. Nature 414(6859):105–111
9. Delaney G, Jacob S, Featherstone C, Barton M (2005) The role of radiotherapy in cancer treatment. Cancer 104(6):1129–1137
10. Ramalho-Santos M, Yoon S, Matsuzaki Y, Mulligan RC, Melton DA (2002) "Stemness": transcriptional profiling of embryonic and adult stem cells. Science 298(5593):597–600
11. Zipori D (2004) The nature of stem cells: state rather than entity. Nat Rev Genet 5(11):1471
12. Gil J, Stembalska A, Pesz KA, Sasiadek MM (2008) Cancer stem cells: the theory and perspectives in cancer therapy. J Appl Genet 49(2):193
13. Singh A, Settleman J (2010) EMT, cancer stem cells and drug resistance: an emerging axis of evil in the war on cancer. Oncogene 29(34):4741–4751
14. Lu B, Chiou SH, Deutsch E, Lorico A (2011) Cancer stem cells. J Oncol 2011, 269437. doi:10.1155/2011/269437, Accessed 7 Jul 2013
15. Gugjoo MB, Amarpal S, Ahmad SR, Yatoo MA, Ahsan-ul-Haq S (2012) Cancer stem cell. Int J Livest Res 2(3):215–218
16. Williams JL (2012) Cancer stem cells. Clin Lab Sci 25(1):50
17. Gupta PB, Onder TT, Jiang G, Tao K, Kuperwasser C, Weinberg R et al (2009) Identification of selective inhibitors of cancer stem cells by high-throughput screening. Cell 138(4):645–659
18. Gupta PB, Chaffer CL, Weinberg RA (2009) Cancer stem cells: mirage or reality? Nat Med 15(9):1010–1012
19. Diehn M, Cho RW, Lobo NA, Kalisky T, Dorie MJ, Kulp AN et al (2009) Association of reactive oxygen species levels and radioresistance in cancer stem cells. Nature 458(7239):780–783
20. Singh A, Greninger P, Rhodes D, Koopman L, Violette S, Bardeesy N et al (2009) A gene expression signature associated with "K-Ras addiction" reveals regulators of EMT and tumor cell survival. Cancer Cell 15(6):489–500
21. Rosen JM, Jordan CT (2009) The increasing complexity of the cancer stem cell paradigm. Science 324(5935):1670–1673
22. Shackleton M, Quintana E, Fearon ER, Morrison SJ (2009) Heterogeneity in cancer: cancer stem cells versus clonal evolution. Cell 138(5):822–829
23. Hirsch HA, Iliopoulos D, Tsichlis PN, Struhl K (2009) Metformin selectively targets cancer stem cells, and acts together with chemotherapy to block tumor growth and prolong remission. Cancer Res 69(19):7507–7511
24. Jamieson CHM, Ailles L, Reya T, Weissman IL. US Patent No. 7,622,255. 24 Nov 2009

25. Liu C, Kelnar K, Liu B, Chen X, Calhoun-Davis T, Li H et al (2011) The microRNA miR-34a inhibits prostate cancer stem cells and metastasis by directly repressing CD44. Nat Med 17(2):211–215
26. Li C, Heidt DG, Dalerba P, Burant CF, Zhang L, Adsay V et al (2007) Identification of pancreatic cancer stem cells. Cancer Res 67(3):1030–1037
27. Li C, Lee CJ, Simeone DM (2009) Identification of human pancreatic cancer stem cells. In: Yu JS (ed) Cancer stem cells. Humana, New York, pp 161–173
28. Al-Hajj M, Wicha MS, Benito-Hernandez A, Morrison SJ, Clarke MF (2003) Prospective identification of tumorigenic breast cancer cells. Proc Natl Acad Sci U S A 100(7):3983
29. Lapidot T, Sirard C, Vormoor J, Murdoch B, Hoang T, Caceres-Cortes J et al (1994) A cell initiating human acute myeloid leukaemia after transplantation into SCID mice. Nature 367:645–648
30. Shimono Y, Zabala M, Cho RW, Lobo N, Dalerba P, Qian D et al (2009) Downregulation of miRNA-200c links breast cancer stem cells with normal stem cells. Cell 138(3):592–603
31. Charafe-Jauffret E, Ginestier C, Iovino F, Wicinski J, Cervera N, Finetti P et al (2009) Breast cancer cell lines contain functional cancer stem cells with metastatic capacity and a distinct molecular signature. Cancer Res 69(4):1302–1313
32. Ginestier C, Liu S, Diebel ME, Korkaya H, Luo M, Brown M et al (2010) CXCR1 blockade selectively targets human breast cancer stem cells in vitro and in xenografts. J Clin Invest 120(2):485
33. Neve RM, Chin K, Fridlyand J, Yeh J, Baehner FL, Fevr T et al (2006) A collection of breast cancer cell lines for the study of functionally distinct cancer subtypes. Cancer Cell 10(6):515–527
34. Zhao C, Chen A, Jamieson CH, Fereshteh M, Abrahamsson A, Blum J et al (2009) Hedgehog signalling is essential for maintenance of cancer stem cells in myeloid leukaemia. Nature 458(7239):776–779
35. Collins AT, Berry PA, Hyde C, Stower MJ, Maitland NJ (2005) Prospective identification of tumorigenic prostate cancer stem cells. Cancer Res 65(23):10946–10951
36. Jachetti E, Mazzoleni S, Grioni M, Ricupito A, Brambillasca C, Generoso L et al (2013) Prostate cancer stem cells are targets of both innate and adaptive immunity and elicit tumor-specific immune responses. Oncoimmunology 2(5), e24520
37. Singh SK, Clarke ID, Terasaki M, Bonn VE, Hawkins C, Squire J et al (2003) Identification of a cancer stem cell in human brain tumors. Cancer Res 63(18):5821–5828
38. Vermeulen L, Felipe de Sousa EM, Van der Heijden M, Cameron K, de Jong JH, Borovski T et al (2010) Wnt activity defines colon cancer stem cells and is regulated by the microenvironment. Nat Cell Biol 12(5):468–476
39. Puglisi MA, Tesori V, Lattanzi W, Gasbarrini GB, Gasbarrini A (2013) Colon cancer stem cells: controversies and perspectives. World J Gastroenterol 19(20):2997
40. Shukrun R, Shakked NP, Dekel B (2013) Targeted therapy aimed at cancer stem cells: Wilm's tumor as an example. Pediatr Nephrol 28(6):1–9
41. Fisher B, Wolmark N, Rockette H, Redmond C, Deutsch M, Wickerham DL et al (1988) Postoperative adjuvant chemotherapy or radiation therapy for rectal cancer: results from NSABP protocol R-011. J Natl Cancer Inst 80(1):21–29
42. Brannon-Peppas L, Blanchette JO (2004) Nanoparticle and targeted systems for cancer therapy. Adv Drug Deliv Rev 56(11):1649
43. Davis ME, Chen Z, Shin DM (2008) Nanoparticle therapeutics: an emerging treatment modality for cancer. Nat Rev Drug Discov 7(9):771
44. Gaitanis A, Staal S (2010) Liposomal doxorubicin and nab-paclitaxel: nanoparticle cancer chemotherapy in current clinical use. Cancer nanotechnology. Humana, New York, pp 385–392
45. Graishar WJ, Tjulndin S, Davidson N, Shaw H, Desai N, Bhar P et al (2005) Phase III trial of nanoparticle albumin-bound paclitaxel compared with polyethylated castor oil–based paclitaxel in women with breast cancer. J Clin Oncol 23(31):7794–7803
46. Farokhzad OC, Jon S, Khademhosseini A, Tran TNT, LaVan DA, Langer R (2004) Nanoparticle-aptamer bioconjugates a new approach for targeting prostate cancer cells. Cancer Res 64(21):7668–7672

47. Kirson ED, Gurvich Z, Schneiderman R, Dekel E, Itzhaki A, Wasserman Y et al (2004) Disruption of cancer cell replication by alternating electric fields. Cancer Res 64(9):3288–3295
48. Stupp R, Wong ET, Kanner AA, Steinberg D, Engelhard H, Heidecke V et al (2012) NovoTTF-100A versus physician's choice chemotherapy in recurrent glioblastoma: a randomised phase III trial of a novel treatment modality. Eur J Cancer 48(14):2192–2202
49. Dylla SJ, Park I-K, Gurney AL (2009) Cancer stem cells. In: Uma L, Chesnut JD, Bhaskar T (eds) Emerging technology platforms for stem cells. Wiley, Hoboken, p 129
50. Pavlovic M, Balint B (2013) Stem cells and tissue engineering. Springer, New York
51. http://www.ncbi.nlm.nih.gov/pmc/articles/PMC3220457/
52. http://ucsdnews.ucsd.edu/pressrelease/monoclonal_antibody_targets_kills_leukemia_cells
53. http://www.ncbi.nlm.nih.gov/pubmed/16998484
54. http://cancerres.aacrjournals.org/content/early/2013/01/04/0008-5472.CAN-12-2221.full.pdf
55. http://www.cancerletters.info/article/S0304-3835%2813%2900832-X/abstract
56. http://clincancerres.aacrjournals.org/content/17/19/6174.full.pdf
57. http://www.ncbi.nlm.nih.gov/pmc/articles/PMC2361410/#bib21
58. www.researchgate.net/...Integrin_alpha_6...cells/.../d912f50.
59. http://www.ncbi.nlm.nih.gov/pmc/articles/PMC3772775/
60. http://www.ncbi.nlm.nih.gov/pmc/articles/PMC3365812/
61. http://mct.aacrjournals.org/content/3/3/345.long
62. http://www.ncbi.nlm.nih.gov/pmc/articles/PMC3730396/
63. http://www.fasebj.org/content/17/9/984.full
64. http://www.jci.org/articles/view/42550

Chapter 5
Genetic Mechanisms Involved in Cancer Stem Cell Emergence

If you wish to make an apple pie from scratch, you must first invent the universe.

Carl Sagan, Cosmos

Current hypotheses suggest that tumors originate from cells that carry out a process of "malignant reprogramming" driven by genetic and epigenetic alterations [1]. The existence of stem-cell-like cells that acquire the ability to self-renew and are able to generate the bulk of more differentiated cells that form the tumor has been confirmed through multiple studies. This population of cancer cells, called cancer stem cells (CSC), is responsible for sustaining the tumor growth and, under determined conditions, can disseminate and migrate to give rise to secondary tumors or metastases to distant organs [1, 2]. Furthermore, CSCs have shown to be more resistant to antitumor treatments than the non-stem cancer cells, suggesting that surviving CSCs could be responsible for tumor relapse after therapy [2]. These important properties have raised the interest in understanding the mechanisms that govern the generation and maintenance of this special population of cells, considered to lie behind the on/off switches of gene expression patterns (Fig. 5.1).

Very recently, by tracking the origin and development of cancer-driving mutations in patients with myelodysplastic syndrome (MDS) the group from Karolinska Institute has offered the conclusive evidence on the existence of CSCs [3]. They used genetic tools to determine in which cells cancer-driving mutations originated and then, propagated into other cancer cells. There was a distinct and rare subset of MDS cells showing all the hallmarks of CSCs including unique tumor propagation. This has already raised the possibilities for antitumor-drug design [4]. However, although most of the CSC markers for different tumors have been discovered, number and function of stem cells in other cancers are likely to differ from that of MDS. The genetics of cancer is a broad field that we shall not consider here in detail.

© The Author(s) 2015
M. Pavlovic, B. Balint, *Bioengineering and Cancer Stem Cell Concept*,
DOI 10.1007/978-3-319-25670-2_5

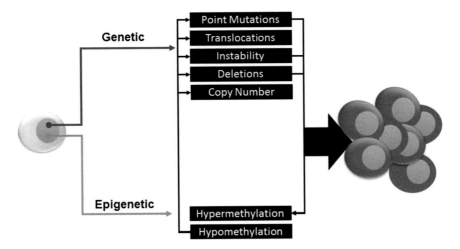

Fig. 5.1 A mechanism-based model of the pathogenesis of human cancer

References

1. Munoz P, Iliou MS, Esteller M (2012) Epigenetic alteration involved in cancer stem cell reprogramming. Mol Oncol 6(6):620–636
2. Esteller M (2012) Cancer, epigenetics and the Nobel Prizes. Mol Oncol 6(6):565–566
3. Woll PS, Kjällquist U, Chowdhury O et al (2014) Myelodysplastic syndromes are propagated by rare and distinct human cancer stem cells *in vivo*. Cancer Cell 25(6):794–808. doi:10.1016/j.ccr.2014.03.036
4. Pabst C, Krosl J, Fares I, Boucher G, Ruel R, Marinier A, Lemieux S, Hébert J, Sauvageau G (2014) Identification of small molecules that support human leukemia stem cell activity ex vivo. Nat Methods 11(4):436–442. doi:10.1038/nmeth.2847

Chapter 6
Epigenetic Mechanisms Involved in Cancer Stem Cell Profiles

For me, I am driven by two main philosophies: know more today about the world than I knew yesterday and lessen the suffering of others. You'd be surprised how far that gets you.

Neil deGrasse Tyson

Introduction

DNA methylation in cancer stem cells: review emphasizing the role of bioinformatics as the aspect of bioengineering

Currently, the role of epigenetic modifications in the development of cancer is an area of research interest within epigenetic regulation processes, important also for normal cell differentiation [1]. Developing research shows evidence that links abnormal patterns of epigenetic modifications with tumorigenesis i.e., emergence and maintenance of cancer cells. Two basic epigenetic mechanisms observed in mammals are DNA methylation and histone modification (Fig. 6.1). DNA methylation may play a significant role in carcinogenesis by silencing genes that are typically unmethylated, e.g., tumor-suppressor genes (TSGs) and microRNAs [2]. Histone modifications are involved in transcription, replication, and repair mechanisms of DNA. Altered histone acetyltransferase (HAT) or histone deacetylase (HDAC) activities have been observed in several types of cancer. Histone modification will not be covered extensively in this chapter.

Beside histones and miRNAs, the methylation of TSGs especially those linked to WNT/β-Catenin, Notch, and Hedgehog signaling pathways are confirmed to be involved in CSC emergence [3]. A very good overview with tables summarizing frequently hypermethylated TSGs contributing to the aberrant activation of WNT/β-Catenin signaling in different CSCs, plus the inhibitors used in cancer treatment clinical trials are given in an article recently published by Sun et al. in 2013 [3]. These data emphasize the need for further understanding of the CSC concept, which has been accepted within the last few years.

DNA methylation is a covalent chemical modification, resulting in the addition of a methyl (CH_3-) group at the carbon 5 position of the cytosine ring. Methylation alters DNA transcription by modifying surrounding chromatin and by inhibiting gene expression through modification of cytosine basis [4]. A vast number of DNA methylation studies have provided a large amount of data potentially useful for the

M. Pavlovic, B. Balint, *Bioengineering and Cancer Stem Cell Concept*,
DOI 10.1007/978-3-319-25670-2_6

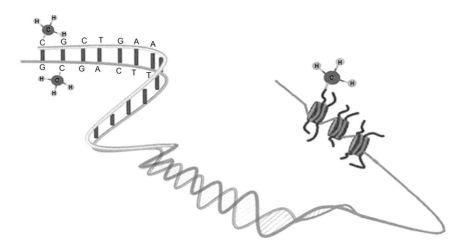

Fig. 6.1 DNA methylation and histone modification

prediction of human cancer [3]. The extent of epigenetic changes to promote carcinogenesis appears to correlate with the level of DNA methylation. Hyper/hypo methylation in chromosomes is associated with abnormal function of DNA methyl transferase (DNMT) enzyme complex, leading to gene imprinting disorders and malignancy [5, 6]. "Hypomethylation usually involves repeated DNA sequences, such as long interspersed nuclear elements, whereas hypermethylation involves CpG islands" [7, 8]. CpG islands are regions of the genome that have a high content of CpG dinucleotides. In normal cells, they are characterized by the absence of DNA methylation and frequent promoter association. This being the case, what kind of aberrant DNA methylation is associated with cancer cells and/or cancer stem cells?

Gene silencing in cancer cells may have its local origin in stem cells [9]. Recent studies suggest that abnormal DNA methylation changes may arise in stem cell compartments inducing cancer development. Thus, the level of aberrant DNA methylation may correspondingly affect the homeostatic balance of the stem cell and turn it into a cancer stem cell [10, 11]. The highly precise mechanisms of altered DNA methylation, which would lead to transformation from normal to cancer stem cell, have not been fully elucidated. However, Zhuang et al. demonstrated that hypermethylation of stem cells can be detected up to 3 years in advance of the appearance of precancerous cells [12]. Recently, a specific mechanism regarding change in methylation state and the ultimate faith of the stem cell in conversion of embryonic stem to mesenchymal-like transition has been identified [13]. There was a genome-wide remodeling of the H3K9 and H3K36 methylation status, promoting this status, which is reversible by addition of ascorbic acid (VC). It is suggested that the L-Pro signaling molecule induction is the promoter of this transition, therefore indicating that this mechanism could serve as a microenvironmental cue for controlling stem cell behavior [13].

These findings suggest that the development of computational methods and modeling techniques that mimic the dynamics of DNA methylation in stem cells in early development of tumorigenesis might be useful in the identification/detection of disrupted methylation patterns that can lead to carcinogenic diseases.

Background and Significance

In the 1940s, Conrad Hal Waddington used the term "epigenetics" for the first time to describe events that could not be explained by classic Mendelian genetic principles [14]. Over the years, the meaning of epigenetics has changed. Epigenetic changes influence gene expression without affecting DNA itself, i.e., epigenetic changes regulate DNA at the transcription level [15–17]. Over the past decade, there has been mounting evidence associating epigenetic modifications to human conditions and diseases such as aging, cancer, lupus, and schizophrenia, viral and cardiovascular diseases [15]. For example, in 1979 Holliday proposed the connection between cancer and inherited epigenetic changes [18]. He suggested that damage to DNA triggers epigenetic changes in gene expression, which are responsible for malignancy [18]. It is also accepted that tumor cells have aberrant patterns of DNA methylation that are often associated with the silencing of tumor-suppressor genes and activation of oncogenes [13]. This leads to changes in signaling pathways, which can also contribute to emergence of malignancy. Birdwell et al. (2014) demonstrated that Epstein-Barr virus (EBV) infection of oral keratinocytes induces CpG island hypermethylation patterns similar to the "CpG island methylator phenotype (CIMP) observed in EBV-associated carcinomas" [19].

There are various mechanisms involved in the epigenetic process. For the purpose of this discussion, only two of those mechanisms will be referenced: DNA methylation and covalent modification of histones [19, 20]. DNA methylation is associated with histone modifications and the interplay of these epigenetic modifications is crucial to regulate the functioning of the genome by changing chromatin architecture [21].

Up today, DNA methylation remains the best-studied posttranscriptional, epigenetic mechanism [22]. During the 1970s, a number of studies have shown the impact of DNA methylation in gene expression and cellular differentiation. Although the exact role of methylation in gene expression is not clear, researchers know that it is essential for cell differentiation and embryonic development [23]. As mentioned, DNA methylation is a biochemical reaction, resulting in the addition of a methyl (CH_3-) group to the cytosine ring at the carbon $5'$ position [7]. This process is catalyzed by enzymes called DNA methyltransferases (DNMTs) and the result is a new base called 5-methylcytosine (5mC) [24, 25].

In mammals there are three main DNA methyltransferases: DNMT1, DNMT3, and DNMT3B. Aberrant methylation patterns seem to be implicated in cancer [24–26]. In cancer cells, TSGs are often silenced due to methylation of the CpG islands. In normal somatic cells, these CpG islands are usually unmethylated, except for some specific regions. Evidence shows that hypermethylation of these regions results in a loss of gene expression by certain genes involved in the development of breast cancer, leukemia, and other hematological diseases. Some of the hypermethylated genes in breast cancer are for example, p16NK4A and estrogen receptor (ER) alpha [7]. Similarly, hypomethylation, which occurs in the context of CpG dinucleotide, is another form of an aberrant methylation pattern. Recent studies confirm that both DNA hypomethylation and DNA hypermethylation are correlated events in cancer development [26]. Furthermore, hypomethylation is a marker of poor prognosis during the development of all four major women-specific cancers: breast, ovarian, endometrial, and cervical cancer [12].

The development of cellular reprogramming technologies has served to better understand cellular lineage determination and can be used to analyze the impact of epigenetic regulations in cancer cells [27–31]. The initial breakthrough in this area was in 1962 by Gurdon and Yamanaka (Nobel Prize in Physiology or Medicine 2012). Three different approaches have been used to induce reprogramming: Cell Fusion, Nuclear Transfers, and induced Pluripotent Stem Cell (iPSC) Technology.

"The goal of reprogramming experimentation has been to generate pluripotent cells from differentiated cells" [29]. Recent studies have shown that the use of reprogramming technologies causes cancer cells to lose tumorigenicity in chronic myeloid leukemia, melanoma, and gastrointestinal cancer cells [28]. These results suggest that nuclear reprogramming may be a therapeutic strategy for the treatment of cancer [30, 31]. However, these studies also suggest that reprogramming technology is not very efficient; cancer cells are resistant to reprogramming and this resistance might be related to the role of epigenetic regulations during reprogramming. Then again, the fact that transformation of iPSCs is accomplished by erasing the epigenetic modification, which is similar to epigenetic modifications found in early stage embryos, demonstrates the significance of epigenetic changes for successful reprogramming, and thus, its role in carcinogenesis [28].

The development of computational methods and modeling techniques may help understand the dynamics between epigenetics and the early stages of cellular reprogramming. Computational methods can also be used to identify critical components that control the DNA methylation machinery and the application of these components as biomarkers for the early diagnosis of human cancers [32]. Although different techniques have been proposed, the complex interplay between the genetic and epigenetic abnormalities found in cancer cells could make it difficult to produce an effective computational model that can predict methylation patterns in different stages of cancer development.

Genetic and Epigenetic Phenomenon in Cancer Development

Cancer is the result of both genetic and epigenetic aberrations [24, 33]. Traditionally, cancer has been considered a genetic disease, but new evidence indicates that epigenetic modifications play significant roles in neoplastic development [24, 34].

Thus far, genetics alone has failed to explain by itself the complex series of events involved in cancer development. Today, advances in research demonstrate the importance of epigenetic processes and indicate that epigenetic changes collaborate with genetic mutations in all stages of cancer development. Epigenetic mechanisms are mitotically heritable and if combined with genetic alterations, may result in the development of cancer [34, 35]. However, unlike genetic mutations, epigenetic changes are reversible [7]. This notion has influenced the development of new therapeutic approaches aimed at reverting gene silencing [35].

Cancer Stem Cell Concept

Carcinogenesis, according to the cancer stem cell theory, is characterized by a multistage process where genetic and epigenetic changes may induce alterations in stem cell function that eventually lead to malignancy. Abnormal gene expression in cancer stem cells (CSCs) is associated with genetic and epigenetic deregulation of important signaling pathways controlling stem cell maintenance, self-renewal, and pluripotency. Some known pathways are WNT/β-Catenin, TGF-β, Hedgehog, and MYC10. Understanding the mechanisms that power epigenome deregulation and malignant stem cell formation could lead to the creation of novel therapeutic technologies.

CSCs can be defined as a population of cells present in tumors that can undergo self-renewal, and differentiation [36]. The abnormal differentiation rate present in cancer cells, the self-renewal capability, and the resistance of various types of cancer to therapy led to the hypothesis that stem cells could induce cancer [19].

Newly minted forms of therapeutic intervention have taken root, and the more subtle role of stem cells is slowly making headway in research. The resultant literature identifies the cancer stem cell as the problem rather than an accepted entity. The concept of CSC should be considered as one that applies only to a subset within the entire spectrum of cancers. Another issue that arises is the ability of cancer to become resistant to targeted treatments due to epigenetic and other mechanisms.

Cancer stem cells, much as healthy totipotent, omnipotent, pluripotent or stem cells of germ tissues, do form the "building blocks" of the resultant tissue. Some of the characteristics of these cells are the ability of self-renewal, differentiation through proliferation, and plasticity. All of these cells are described as possessing "stemness," a general term referring to two main characteristics: phenotypic characteristic of particular cancer stem cells (existence of tumor-specific molecular markers) and tumorigenic functionality, e.g., capability to develop into a growing cancer. Therefore, the ideal model of cancer therapy is one where these cancer stem cells are targeted specifically, then destroyed, thereby causing the tumor's mass to also die by removing its source of viability and regeneration. With as great specificity as possible, this model promises little-to-no side effects versus current treatment regimens. However, an accurate cancer stem cell model is necessary in order to create an appropriate and targeted design. How can we do that?

Cancer Stem Cell Models: Theories and Possible Explanations

Despite numerous studies that support the CSC hypothesis, there is still a lot of controversy surrounding the idea of an acceptable cancer stem cell model that can predict tumor formation. Two models of carcinogenesis have been largely discussed: the stochastic model (genetic mutation model) and the hierarchic model (cell-of-origin model). The stochastic model is based on the interpretation that tumors result from random mutations in any normal cell. The hierarchic model, also known as CSC model (Fig. 6.2), proposes that cancer initiates from a single CSC (altered normal stem cell) which gives rise to a hierarchical organization and sequential differentiation of cells such that only a small group of these cells are responsible for tumor initiation and maintenance. Quite recently, these investigators proposed that one of the mechanisms for heterogeneous cancer stem cell population is tumor cell oncogene addiction. There are two aspects to this addiction hypothesis. First, the maintenance of the tumor phenotype depends on the continued expression of certain oncogenes. Second, the expression of certain lineage pathways in the cell is dependent on (i.e., addicted to) the expression of those oncogenes, which contributes to further heterogeneity [37–39]. In addition to this, epigenetic modifications are thought to be involved in cell differentiation. Studies in cancer-associated fibroblasts displaying DNA methylation and chromatin modification patterns imply evidence of cellular heterogeneity in the stroma (Fig. 6.3) [37].

Data obtained in various types of melanoma tumors and BCR-ABL1 lymphoblastic leukemia have shown that not all of the tumors existing in the human population can be explained by the CSC hierarchical organization theory [38]. This is because the linear-hierarchical organization defined by the CSC hypothesis is

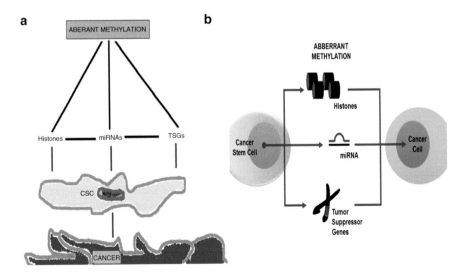

Fig. 6.2 (**a**) and (**b**). Epigenetic DNA methylated modifications

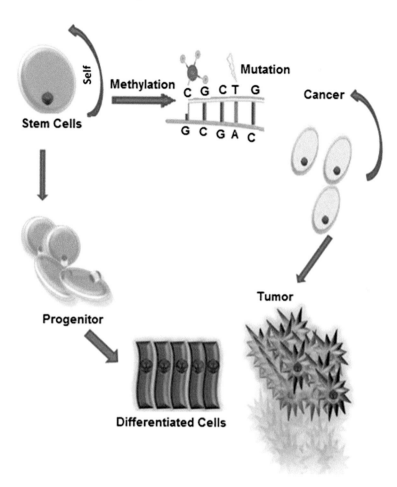

Fig. 6.3 Cancer stem cell model

not a consistent feature of tumor behavior. However, an increasing amount of data shows that the CSC hypothesis can be adequate to explain some tumor's growth and their heterogeneity. This is also in agreement with the view that not all the tumors originate/emerge from the cancer stem cells (as holds true for tumors of viral origin).

Cancer Stem Cell Markers: Conceptual Features

"Cancers can harbor heterogeneous and biologically distinct populations of CSCs" [38]. Most of the research data have shown that DNA methylation is an essential epigenetic mechanism to protect stem cells from premature activation of predominant differentiation programs. The data also suggests that methylation dynamics determine stem cell functions in tissue homeostasis and cancer. Using in vivo and in vitro experimental methods, CSCs have been identified by isolating cancer cells with a unique surface molecular expression profile within tumor cells known as a phenotype [39]. Among the most common CSC markers are some cluster destination antigens (CDs: CD133, CD44, CD166) or other molecules such as: ALDh1, EpCAM, etc [40]. Tumor cells expressing these markers have shown to have stem cell characteristics. However, the presence of these markers is not universal for all cancer types and they are not exclusively expressed by CSCs [38–40]. Stem cell markers cannot be considered a dependable biomarker because studies have shown that the presence of particular markers can vary between patients with the same tumor type. However, markers like CD44, CD24, CD90, ALDH, etc. have been proposed to identify a distinctive group of cells with a self-renewal capacity in a variety of epithelial tumors [40].

The stem cell marker concept suggests the marker does not change during CSC division because it is in a renewal, meaning the marker perpetuates its same phenotype during its fast and highly repetitive clonal expansion. This stability of the protein marker during self-renewal of CSC and phenotype propagation enables us to reliably identify cancer stem cells by in vivo and in vitro assays. Further studies may bring new advances in this subject (Table 6.1).

Significance of Computational Approaches for Cancer Prediction: Identifying Methylation Profiles in Cancer Stem Cells

"Abnormal DNA methylation and its subsequent effects on gene activity are a hallmark of cancer" [12]. Several studies have confirmed that DNA methylation patterns are not random in normal cells [41–55]. Most importantly, Zhuang et al. (2012) confirmed those patterns could be detected previous to the appearance of precancerous cells [12]. The significance of these findings represents an important contribution to cancer research, because it indicates the possible early intervention through the identification of an early cancer-leading signature. Computational methods and modeling techniques may help uncover the molecular mechanisms that trigger human cancers and formation of cancer stem cells.

Some recent research on methylation and pathways in cancer stem cells of colorectal cancer and tissues predict poor prognosis [55]. For example, aberrant methylation may reflect a differentiation status of the malignant disease. However, West et al. (2013) have found age/associated differential methylation occurring in stem cell differentiation pathways [56]. They used a network algorithm to identify tissue-independent, age/associated "hot spots" targeting these pathways. Since age increases the risk for cancer, it may also be instructive to investigate pathways associated with aberrant methylation related to aging and the possible involvement of these pathways to aberrant methylation occurring in CSCs [52].

Table 6.1 Cancer stem cell properties, assays, drugs in use, markers (compilation from many sources)

Property	Definition/origin	Marker CSC	Signaling pathway	Assay: in vivo/in vitro	Inhibitors	Reference
Self-renewal	Stochastic model	Depends on the type of CSC	Depends on the type of CSC	Serial transplantations in NOD/SCID mice Re-plating assays	Depends on the type of CSC	Many sources
Limited differentiation capacity	Hierarchical model	AML (CD34+/CD38−)	Wnt	Differentiation assays in vitro Transplantation		Bonnet and Dick. Nat Med 1973; 73
Tumor initiation/metastases	Ability to propagate tumor when transplanted in proper environment	Pancreas	Notch	Sphere formation Invasion assays Transplantation		Different sources
Relapse	The property of resistance to different therapies and the ability to re-occur	Ovarian Breast Colorectal Wilm's tumor	JAK-STAT, Hedgehog	Chemo/radio-resistance assays		Different sources

The CD133 gene silencing was found to be associated with hypermethylation at the promoter region in the CpG island in colon cancer and demethylated with 5-aza-2′-deoxycytidine (DAC) (Mi et al. in 2008) [57]. This suggests that DNA methylation is a mechanism for CD133 transcriptional inactivation in colon cancer [41]. Transcriptional silencing of the CD44, CD34, and EpCAM genes in colon cancer has also been identified [41]. Although there is no evidence that CSC markers have any role in CSC functioning, its correlation with cancer stem cells and tissues, as previously mentioned, predicts poor prognosis.

Since methylation seems to be an early indicator of cancerogenesis and can be detected in body fluids, it may be of potential use in early detection of tumors and thereby improving the prognosis of cancers. The rapid advancement of new technologies and the expansion of a bioinformatics infrastructure have made possible the mapping of DNA methylation markers in the genome on a large scale.

Computational models targeting methylated or unmethylated CpG dinucleotide and CpG islands have been developed to complement the wet chemistry approach [42]. Although great progress has been made, so far there is not an established DNA methylation signature that can be used in the early diagnosis or prognosis across multiple human cancers [3, 12]. The dynamics of DNA methylation is complex considering all the biological processes involved along with its relationship to transcriptional control. However, the development of computational techniques presents a new dimension in discovering cancer biomarkers and potentially predicting cancer outcome based on epigenetic mechanisms.

It is crucial to validate the role of aberrant DNA methylation in cancer stem cells. In recent years several methods have been developed to provide a genomewide picture of the state of DNA methylation, including: nextgeneration genomewide sequencing of bisulfiteconverted DNA; methylated DNA immunoprecipitation (MeDIP) followed by either hybridization to highdensity oligonucleotide arrays or nextgeneration sequencing; and dedicated Illumina 27 and 450 K arrays that measure the state of methylation of well-characterized CG sites distributed in the genome. Although genome-wide sequencing is still prohibitively costly for large population studies, array approaches are being frequently used to delineate DNA methylation signatures of disease states in primary clinical material rather than cell lines.

An excellent review on analyzing and interpreting methylation data in a broad range of biological processes and diseases is given recently by Bock (Bock 2012) [58]. It can serve as a guide to approaches for answering critical questions regarding the significance of altered DNA methylation in particular cancer stem cells, as we get more specific information in future research. We still do not know whether altered DNA methylation is the cause or consequence of the malignancy and what is the entire spectrum of possible aberrant events. Thus, we propose the following workflow in wet-lab (assays for DNA methylation mapping) and computational approach that could be used to identify DNA methylation patterns and their significance in cancer stem cells (Fig. 6.4).

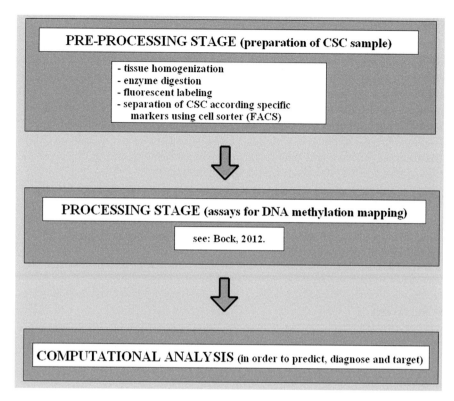

Fig. 6.4 Proposed diagram for sequence of events in wet-lab and computational approach in cancer stem cell DNA methylation processes validation

The following computational methodologies have been applied or combined in order to aid in the identification/discovery of cancer stem cell markers, which may lead to a more complete understanding of the cancer stem cell model:

1. The method for prediction of DNA methylation in CpG dinucleotides using support vector machine (SVM) [44].
2. The linear discriminant analysis method [45].
3. The artificial neural networks (ANN) [46].
4. The hidden Markov models (HMM) [47].
5. The naïve Bayes method [48], logistic regression method [49].
6. Probabilistic Bayesian Networks [50].

Novel artificial intelligence and machine learning algorithms are being continually developed in data mining research and may be applied to cancer stem cell biomarker discovery. These algorithms might prove even better for discrimination of potential biomarkers. In addition, "a centralized repository for epigenetic-related data is still lacking" [51]. The establishment of such a repository could significantly accelerate the research which have shown already strong expansion [58–67].

Conclusions

DNA methylation plays a significant role in gene expression. Aberrant methylation changes have been associated with tumor development. Understanding the mechanisms involved in the epigenetic machinery can identify potential clinical applications in cancer diagnosis, prognosis, and new therapeutic approaches. Since there are not extensive computational approaches developed in this field, the development of computational models and bioinformatics tools in addition to wet chemistry advances could potentially improve our understanding of epigenetic modifications and its association with human cancers.

As the ultimate goal, the potential for targeting cancer stem cells specifically in therapy models is proving to be promising. It is anticipated that this direction will be continued in conjunction with other current treatment modalities and will be manipulated to have greater cancer regression rates and better prognoses for the patients suffering these neoplasms.

References

1. Richly HL, Aloia L, Di Croce L (2011) Roles of the polycomb group proteins in stem cells and cancer. Cell Death Dis 2(9), e204
2. Pecorino L (2008) Evidence of a role for epigenetics in carcinogenesis. In: Pecorino L (ed) Molecular biology of cancer, 2nd edn, vol 3. Oxford University Press, New York, p 58
3. Sun L, Yang A, Cabarcas SM, Mathews LA, Farrar WL (2013) Perspectives and challenges in epigenetic therapy: DNA methylation in cancer stem cells. In: Tatarinova TV, Sablok G (eds) DNA methylation. Nova Science, New York. ISBN 978-1-62417-128-4
4. Agarwal A, Polineni R, Hussein Z, Vigoda I, Tushar D, Bhagat SB et al (2012) Role of epigenetic alterations in the pathogenesis of Barrett's esophagus and esophageal adenocarcinoma. Int J Clin Exp Pathol 5(5):382–396
5. Raghavan K, Ruskin HJ (2012) Modeling DNA methylation dynamics. In: Tatarinova T, Kerton O (eds) DNA methylation—from genomics to technology, 1st edn. InTech, Rijeka, pp 3–28
6. Chahwan R, Wontakal SN, Roa S (2011) The multidimensional nature of epigenetic information and its role in disease. Discov Med 11(58):233–243
7. Das P, Rakesh S (2004) DNA methylation and cancer. J Clin Oncol 22:4632–4642
8. Ehrlich M (2002) DNA methylation in cancer: too much, but also too little. Oncogene 21:5400–5413
9. Esteller M (2007) Epigenetic gene silencing in cancer: the DNA hypermethylome. Hum Mol Genet 16(Review Issue 1):R50–R59
10. Marquardt JU, Thorgeirsson SS (2010) Epigenetic regulation of cancer stem cells in liver cancer: current concepts and clinical implications. Laboratory of experimental carcinogenesis. J Hepatol 53(3):568–577
11. Broske AM, Vockentanz L, Kharazi S, Huska MR, Mancini E, Scheller M et al (2009) DNA methylation protects hematopoietic stem cell multipotency from myeloerythroid restriction. Nat Genet 41:1207–1215
12. Zhuang J, Jones A, Lee S-H, Ng E, Fiegl H et al (2012) The dynamics and prognostic potential of DNA methylation changes at stem cell gene loci in women's cancer. PLoS Genet 8(2), e1002517

13. Jablonka E, Marion JL (2002) The changing concept of epigenetics. Ann N Y Acad Sci 981:82–96
14. Goldberg AD, Allis CD, Bernstein E (2007) Epigenetics: a landscape takes shape. Cell 128(4):635–638
15. Stauffer B, DeSouza CA (2010) Epigenetics: an emerging player in health and disease. J Appl Physiol 109(1):230–231
16. Fraga M, Esteller M (2007) Epigenetics and aging: the targets and the marks. Trends Genet 23:413–418
17. Van Vliet J, Oates NA, Whitelaw E (2007) Epigenetic mechanisms in the context of complex diseases. Cell Mol Life Sci 64:1531–1538
18. Holliday R (1979) A new theory of carcinogenesis. Br J Cancer 40(4):513–522
19. Dricu A, Purcaru SO, Buteica AS, Tache DE, Daianu O, Stoleru B et al (2012) DNA methylation, stem cells and cancer, methylation—from DNA, RNA and histones to diseases and treatment. InTech, Rijeka. doi:10.5772/53263. ISBN 978-953-51-0881-8
20. Cedar H, Bergman Y (2009) Linking DNA methylation and histone modification: patterns and paradigms. Nat Rev Genet 10(5):295–304
21. Kulis M, Esteller M (2010) DNA methylation and cancer. Adv Genet 70:27–56
22. Herranz M, Esteller M (2007) DNA methylation and histone modifications in patients with cancer: potential prognostic and therapeutic targets. In: Sioud M (ed) Target discovery and validation reviews and protocols: emerging molecular targets and treatment options, vol 2. Humana, Totowa, pp 225–262
23. Phillips T (2008) The role of methylation in gene expression. Nature Educ 1(1):116, http://www.nature.com/scitable/topicpage/the-role-of-methylation-in-gene-expression-1070
24. Kanwal R, Gupta S (2010) Epigenetic and cancer. J Appl Physiol 109(2):598–605
25. Issa JP, Kantarjian HM (2009) Targeting DNA methylation. Clin Cancer Res 15:3938–3946
26. Ehrlich M (2009) DNA hypomethylation in cancer cells. Epigenomics 1(2):239–259
27. Lensch W, Mummery CL (2013) From stealing fire to cellular reprogramming: a scientific history leading to the 2012 Nobel Prize. Stem Cell Rep 1(1):5–17
28. Semi K, Matsuda Y, Ohnishi Y, Yamada Y (2013) Cellular reprogramming and cancer development. Int J Cancer 132(6):1240–1248
29. De Carvalho D, Jueng S, Jones P (2010) DNA methylation and cellular reprogramming. Trends Cell Biol 20(10):609–617
30. Tafani M, Russo MA (2012) Reprogramming cancer stem cells. J Cancer Sci Ther 4:25–26
31. Zhang X, Cruz FD, Terry M, Remotti F, Matushansky I (2012) Terminal differentiation and loss of tumorigenicity of human cancer via pluripotency-based reprogramming. Oncogene 32:2249–2260
32. Zheng H, Shi-Wen J, Wu H (2011) Enhancement on the predictive power of the prediction model for human genomic DNA methylation. http://www.lidi.info.unlp.edu.ar/WorldComp 2011–Mirror/BIC4342.pdf.
33. Link A, Balaguer F, Goel A (2010) Cancer chemoprevention by dietary polyphenols: promising role for epigenetics. Biochem Pharmacol 80(12):1771–1792
34. Herceg Z (2007) Epigenetics and cancer: towards an evaluation of the impact of environmental and dietary factors. Mutagenesis 22:91–103
35. Jones P, Baylin SB (2007) The epigenomic of cancer. Cell 128:683–692
36. Appalaraju J, Elkord E (2012) Significance of CD44 and CD24 as cancer stem cell markers: an enduring ambiguity. Clin Dev Immunol 2012:1–11
37. Melero-Martin JM, Dudley AC (2011) Concise review: vascular stem cells and tumor angiogenesis. Stem Cells 29(2):163–168
38. Visvader JE, Lindeman GJ (2012) Cancer stem cells: current status and evolving complexities. Cell Stem Cell 10(6):717–728
39. Bonnet D, Dick JE (1997) Human acute myeloid leukemia is organized as a hierarchy that originates from a primitive hematopoietic cell. Nat Med 3:730–737. doi:10.1038/nm0797-730

40. Donnenberg VS, Donnenberg AD, Zimmerlin L, Landreneau RJ, Bhargava R, Wetzel RA, Basse P et al (2010) Localization of CD44 and CD90 positive cells to the invasive front of breast tumors. Cytometry B Clin Cytom 78B(5):287–301

41. Yi JM (2012) DNA methylation of stem cell surface markers in colon cancers. Genes Genomic 34(3):299–304

42. Zheng H, Wu H, Li J, Jiang S-V (2013) CpGMethPred: computational model for predicting methylation status of CpG islands in human genome. BMC Med Genomics 6(Suppl 1):S13

43. Serman A, Vlahovic M, Serman LJ, Bulic-Jakus F (2006) DNA methylation as a regulatory mechanism for gene expression in mammals. Coll Antropol 30:665–671

44. Manoj B, Zhanga H, Reinherza EL, Rechea PA (2005) Prediction of methylated CpGs in DNA sequences using a support vector machine. FEBS Lett 579:4302–4308

45. Wei D et al (2008) Methylation Linear Discriminant Analysis (MLDA) for identifying differentially methylated CpG islands. BMC Bioinformatics 9(1):337

46. Firpi HA, Ucar D, Tan K (2010) Discover regulatory DNA elements using chromatin signatures and artificial neural network. Bioinformatics 26(13):1579–1586

47. Seifert M et al (2012) MeDIP-HMM: genome-wide identification of distinct DNA methylation states from high-density tiling arrays. Bioinformatics 28(22):2930–2939

48. Down TA et al (2008) A Bayesian deconvolution strategy for immunoprecipitation-based DNA methylome analysis. Nat Biotechnol 26(7):779–785

49. Yang Y, Nephew K, Kim S (2012) A novel k-mer mixture logistic regression for methylation susceptibility modeling of CpG dinucleotides in human gene promoters. BMC Bioinformatics 13(Suppl 3):S15

50. Yu H et al (2008) Inferring causal relationships among different histone modifications and gene expression. Genome Res 18(8):1314–1324

51. Lim SJ, Tan TW, Tong JC (2010) Computational epigenetics: the new scientific paradigm. Bioinformation 4(7):331–337

52. Pavlovic M, Balint B (2013) Stem cells and tissue engineering. Springer, New York

53. Robertson DK (2001) DNA methylation, methyltransferases, and cancer. Oncogene 20(24):3139–3155

54. De S, Shaknovich R, Riester M, Elemento O, Geng H et al (2013) Aberration in DNA methylation in B-cell lymphomas has a complex origin and increases with disease severity. PLoS Genet 9(1), e1003137. doi:10.1371/journal.pgen.100313

55. Birdwell CE, Queen KJ, Kilgore P et al (2014) Genome-wide DNA as an epigenetic consequence of Epstein-Barr virus infection of immortalized keratinocytes. J Virol 88:11442–11458. doi:10.1128/JVI.00972-14

56. Munoz P, Iliou MS, Esteller M (2012) Epigenetic alteration involved in cancer stem cell reprogramming. Mol Oncol 6:620–636

57. Esteller M (2012) Editorial: cancer, epigenetics and the Nobel Prizes. Mol Oncol 6:565–566

58. Sinkkonen L, Hugenschmidt T, Berninger P, Gaidatzis D, Mohn F, Artus-Revel CG, Zavolan M, Svoboda P, Filipowicz W (2008) MicroRNAs control de novo DNA methylation through regulation of transcriptional repressors in mouse embryonic stem cells. Nat Struct Mol Biol 15(3):259–267

59. Warlick ED, Cioc A, Defor T, Dolan M, Weisdorf D (2009) Allogeneic stem cell transplantation for adults with myelodysplastic syndromes: importance of pretransplant disease burden. Biol Blood Marrow Transplant 15:30–38

60. Cutler CS, Lee SJ, Greenberg P, Deeg HJ, Perez WS, Anasetti C et al (2004) A decision analysis of allogeneic bone marrow transplantation for the myelodysplastic syndromes: delayed transplantation for low-risk myelodysplasia is associated with improved outcome. Blood 104(2):579–585

61. Jin H, Ma Y, Shen Q, Wang X (2012) Circulating methylated DNA as biomarkers for cancer detection. In: Dricu A (ed) Methylation—from DNA, RNA and histones to diseases and treatment. InTech, Rijeka. ISBN 978-953-51-0881-8

62. Katayama S, Tomaru Y, Kasukawa T, Waki K, Nakanishi M, Nakamura M, Nishida H, Yap CC, Suzuki M, Kawai J, Wahlestedt C (2005) Antisense transcription in the mammalian transcriptome. Science 309:1564–1566

63. Carninci P, Kasukawa T, Katayama S, Gough J, Frith MC, Maeda N, Oyama R, Ravasi T, Lenhard B, Wells C, Wahlestedt C et al (2005) The transcriptional landscape of the mammalian genome. Science 309:1559–1563

64. Wahlestedt C (2013) Targeting long noncoding RNA to therapeutically modulate gene expression. Nat Rev Drug Discov 12:433–446. doi:10.1038/nrd4018

65. Rinn JL, Kertesz M, Wang JK, Squazzo SL, Xu X, Brugmann SA, Goodnough LH, Helms JA, Farnham PJ, Segal E et al (2007) Functional demarcation of active and silent chromatin domains in human HOX loci by noncoding RNAs. Cell 129:1311–1323

66. Gupta RA, Shah N, Wang KC, Kim J, Horlings HM, Wong DJ, Tsai MC, Hung T, Argani P et al (2010) Long non-coding RNA HOTAIR reprograms chromatin state to promote cancer metastasis. Nature 464:1071

67. Pastori C, Penas C, Volmar C-H, Johnstone L, Brothers SP, Graham RM, Bryce A, Sarkaria JN, Komotar RJ, Wahlestedt C, Ayad NG (2014) BET bromodomain proteins are required for glioblastoma cell proliferation. Epigenetics 9(4):611–620

Chapter 7
Mitochondrial Respiration of Cancer Stem Cell

I believe there is another world waiting for us. A better world.
And I'll be waiting for you there.

David Mitchell, Cloud Atlas

Introduction

Like normal stem cells, CSCs self-renew and proliferate and in that way contribute to cancer growth and metastasis [1]. They are considered responsible for relapse of the tumor and therefore, ultimate target of cancer therapy. They are also extremely drug-resistant cells and for that reason, difficult to completely eradicate. Tumorigenesis arises from a complexity of different signals between different effectors. Tumor is regarded as a heterogeneous group of cells, the mitochondria of which present a shift on energetic metabolism with silenced mitochondria and their great impact upon cell fate determination [1–9]. Malignant cell mitochondria are unable to recognize malignant signals and trigger a normal physiological response conferring both the resistance to cell death and a selective advantage. Thus, by silencing catabolic processes (apoptosis, necrosis, and autophagy) cancer cells have remodeled signaling pathways conferring resistance to cell death and increasing proliferative and invasion processes [9]. A very important aspect of future research is to understand how CSCs differ from normal stem cells and differentiated cells, and what is the underlying mechanism that makes them harder to eliminate.

The Role of Mitochondria in Energy Metabolism: Oxidative Phosphorylation and Uncoupling Effect, Apoptosis, Calcium, and ROS Balancing

Mitochondria are energy-producing organelle in the cell, with essential function in cellular respiration through the function of the enzymes of the respiratory chain located in the inner mitochondrial membrane [1–4]. Two Nobel Prizes are awarded for fundamental discoveries of: chemiosmosis (Peter Mitchel) and oxidative

© The Author(s) 2015
M. Pavlovic, B. Balint, *Bioengineering and Cancer Stem Cell Concept*,
DOI 10.1007/978-3-319-25670-2_7

phosphorylation mechanism (rotational catalysis in ATP synthesis—Hans Boyer). The basic source of energy for ATP synthesis is proton–motive force of the respiratory chain, during which three essential events are going on: separation of the charge on the inner mitochondrial membrane (negative inside and positive outside), synthesis of endogenous water, and synthesis of ATP in mitochondria. The last event is performed on the ATP-ases of the inner mitochondrial membrane during rotational catalysis when ADP, being close to inorganic phosphate (Pi) creates the ester bond high in energy. Thus, respiration is the process coupled to oxidative phosphorylation, which is nothing else but synthesis of ATP. The consequence of prevention of the coupling of the respiration with oxidative phosphorylation is known as uncoupling effect [5–7]. It will decrease the energy level of mitochondrial compartment as well as cell in toto, since 95 % of ATP is synthesized on the ATP-ases of the inner mitochondrial membrane and only 5 % on the ATP-ases of the plasma membrane—from cytoplasmic, glycolytic process [5, 6].

The other important role of mitochondria is in apoptosis/programed cell death (PCD) [8, 9]. This event includes the interaction of well-characterized genetic programs with central role to mitochondria. During this process, mitochondrial inner membrane potential decreases, and proteins contained in the intermembrane space including cytochrome c and apoptosis-inducing factor, are released. With the help of enzymes caspases, cysteine proteases, the process will advance toward DNA, blabbing of the membrane and fragmentation of the cell into apoptotic bodies. Thus, two vital processes are governed with the mitochondria in the cell including CSCs. Therefore, we can look into them as targets for CSC therapy.

Mitochondria are also the storage/buffering system for Ca++ and balancing system for ROS production, to minimize oxidative damage. With autonomous, circular mtDNA, it is also organelle capable for division, with possible role in self-renewal potential, and stemness maintenance [8–12]. During early stages it has perinuclear localization in the cell, making crosstalk with nucleus more efficient. Nuclear genome regulates approximately 1500 mitochondria-related genes [7–9].

Recent studies have revealed significant role of mitochondria in the maintenance of pluripotency, differentiation, and reprogramming of IPSCs [1, 2]. It seems that pluripotent states are linked to glycolytic energy production, while cell differentiation requires increase in mitochondrial oxidative phosphorylation [2]. Actually, several studies have shown the necessity for the switch from oxidative phosphorylation to glycolysis to reprogram somatic cell to pluripotency [1, 2]. Thus, Zhang et al. have recently shown that mitochondrial respiration regulates adipogenic differentiation of human mesenchymal stem cells (MSCs) [3].

SC mitochondria play important role in stemness and differentiation [8, 9]. They are also localized in perinuclear space of ES [9]. It is still uncertain whether the role of CSC mitochondria is the same in SCs of different origin or in cancer cells [9]. More studies are necessary to accomplish the judgement of that issue. According to some studies, CSCs mitochondria show different roles and features in the cancers of different origin [9–13]. And although the mitochondrial features of CSCs in several cancers are different, CSCs again differ from those of noncancer stem cells. This indicates the necessity for further research development.

The Change in Lipid Content on the Inner Mitochondrial Membrane and Uncoupling Effect of Mitochondria in the Cancer Cells

In order to determine the role of mitochondria in particular features of CSCs, a lot of studies have been done upon its function and lipid content of the inner mitochondrial membrane. Thus, it is not still quite certain whether mitochondrial inner membrane lipid content is different in CSCs and normal or other non-CSCs [9]. This field requires more efforts in order to be clarified.

However, there are several reports indicating the uncoupling/decoupling of respiration from energy production [5–7]. Deeper research has also shown close interrelationship between respiration and glycolysis in cancer cells, such as Warburg (suppression of glycolysis by respiration) and Crabtree effect (suppression of respiration by glycolysis), which are present in the tumors with high glycolytic activity [14–26]. Besides Warburg and Crabtree, these phenomena are observed, reported, and studied later on, by E. Racker [24, 25]. It is of great importance for the life of CSCs how their metabolism reflects upon the function of mitochondria and vice versa [27–33]. The importance of these phenomena is realized now and further studies will define the specific features of these metabolic and energy conversions in each CSCs type. This will facilitate the choice of the targeted therapy on molecular level and clarify which pathways or molecular structures in mitochondria should be targeted in order to specifically kill/destroy CSCs.

Therapeutic Approaches: Targeting Mitochondria in Cancer Stem Cells

Development of effective therapeutic strategies to eliminate cancer stem cells, which play a major role in drug resistance and disease recurrence, is critical to improve cancer treatment outcomes [9]. In a recent study published in Nature methods (2014), the identification of two new compounds that allow to maintain leukemia cells in culture is described [34]. Growing myeloid cells in culture is a major breakthrough that will enable the study of particular mechanisms that regulate the survival and proliferation of leukemic cells as well as their resistance to cancer drugs [34].

However, a recent study showed that glioblastoma stem cells (GSCs) exhibited low mitochondrial respiration and high glycolytic activity [13]. These GSCs were highly resistant to standard drugs such as carmustine and temozolomide (TMZ), but showed high sensitivity to a glycolytic inhibitor 3-bromo-2-oxopropionate-1-propyl ester (3-BrOP), especially under hypoxic conditions. The research further showed that combination of 3-BrOP with carmustine but not with TMZ achieved a striking synergistic effect and effectively killed GSCs through a rapid depletion of cellular ATP and inhibition of carmustine-induced DNA repair. This drug combination significantly impaired the sphere-forming ability of GSCs in vitro and tumor formation in vivo, leading to increase in the overall survival of mice bearing orthotropic inoculation of GSCs. Further mechanistic study showed that 3-BrOP and carmustine inhibited glyceraldehyde-3-phosphate dehydrogenase and caused a severe energy crisis in GSCs. This study suggests that GSCs are highly glycolytic and that certain drug combination strategies can be used to effectively overcome their drug resistance based on their metabolic properties (Fig. 7.1).

The novel approach is using antibiotics as the tool to eradicate CSCs [11, 12]. Here is the summary on five classes of antibiotics, along with their corresponding molecular targets which all converge on mitochondrial biogenesis or oxidative phosphorylation (OXPHOS).

Thus, there is the hope for both liquid and solid tumors to be treated through mitochondrial and energy metabolism pathways.

Conclusions

Mitochondria is the organelle tied to the nucleus determined by its perinuclear position and crosstalk that reflects interaction. Therefore, it is organelle of major importance to the normal and CSCs. Although there are still a lot of unclarified aspects of

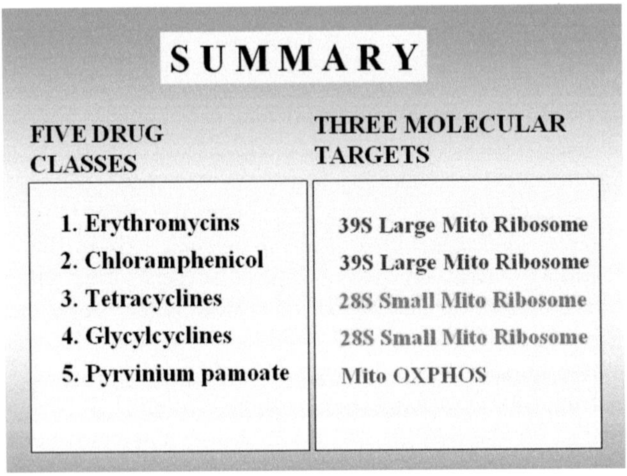

SUMMARY	
FIVE DRUG CLASSES	**THREE MOLECULAR TARGETS**
1. Erythromycins	39S Large Mito Ribosome
2. Chloramphenicol	39S Large Mito Ribosome
3. Tetracyclines	28S Small Mito Ribosome
4. Glycylcyclines	28S Small Mito Ribosome
5. Pyrvinium pamoate	Mito OXPHOS

Fig. 7.1 Antibiotics and molecules that are targeted in mitochondria

its role in CSCs, accumulated data indicate that there could be a way to target CSCs through some of its functions or particular molecules. The target would be in essence:

- Apoptotic pathways and molecules
- Respiration
- Glycolysis
- Uncoupling effect (prevention of binding respiration with OXPHOS, rendering CSCs anergic)
- Antibiotic treatment of molecular targets (39 S Large Mito Ribosome, 28 S small Mito Ribosome and Mito OXPHOS)

The most interesting is the fact the discoveries about mitochondrial significance in the tumorigenesis which have been done hundred years ago mostly, brought up into modern time open the door to rational drug design for CSCs therapy [35–37] (Fig. 7.2).

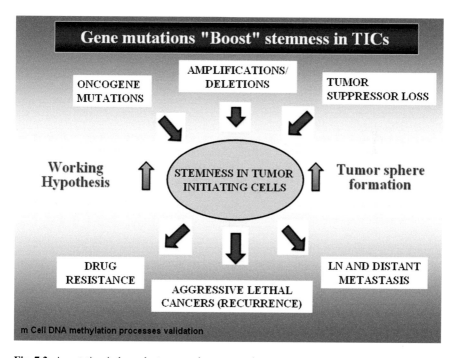

Fig. 7.2 A mutation-independent approach to cancer therapy

References

1. Bukowiecki AJ, Prigione A (2014) Mitochondrial function in pluripotent stem cells and cellular reprogramming. Gerontology 60:174–182. doi:10.1159/000355050
2. Xu X, Duan S, Yi F, Ocampo A, Liu GH, Izpisua Belmonte JC (2013) Mitochondrial regulation in pluripotent stem cells. Cell Metab 18:325–332
3. Zhang Y, Marsboom G, Toth PT, Rehman J (2013) Mitochondrial respiration regulates adipogenic differentiation of human mesenchymal stem cells. PLoS One 8(10):77077
4. Cho YM et al (2006) Dynamic changes in mitochondrial biogenesis and antioxidant enzymes during the spontaneous differentiation of human embryonic stem cells. Biochem Biophys Res Commun 348:1472–1478
5. Couplan E, del Mar Gonzalez-Barroso M, Alves-Guerra MC, Ricquier D, Goubern M, Bouillaud F (2002) No evidence for a basal, retinoic, or superoxide-induced uncoupling activity of the uncoupling protein 2 present in spleen or lung mitochondria. J Biol Chem 277:26268–26275
6. Emre Y, Nubel T (2010) Uncoupling protein UCP2: when mitochondrial activity meets immunity. FEBS Lett 584:1437–1442
7. Shyh-Chang N, Zheng Y, Locasale JW, Cantley LC (2011) Human pluripotent stem cells decouple respiration from energy production. EMBO J 30:4851–4852
8. Facucho-Oliveira JM, Alderson J, Spikings EC, Egginton S, St John JC (2007) Mitochondrial DNA replication during differentiation of murine embryonic stem cells. J Cell Sci 120:4025–4034
9. Song IS, Jeong JY, Jeong SH, Kim HK, Ko KS, Rhee BD, Kim N, Han J (2015) Mitochondria as therapeutic targets for cancer stem cells. World J Stem cells 7(2):418–427
10. Fulda S, Galluzzi L, Kroemer G (2010) Targeting mitochondria for cancer therapy. Nat Rev Drug Discov 9:447–464
11. Loureiro R, Mesquite K, Oliveira PJ, Vega-Naredo I (2013) Mitochondria in cancer stem cells: a target for therapy. Recent Pat Endocr Metab Immune Drug Discov 7(2):102–114
12. Lamb R, Ozsvari B, Lisanti CL, Tanowitz HB, Howell A, Martinez-Outschoorn UE, Sotgia F, Lisanti MP (2015) Antibiotics that target mitochondria effectively eradicate cancer stem cells, across multiple tumor types: treating cancer like an infectious disease. Oncotarget 6(7):4569–4584
13. Folmes CD, Nelson TJ, Martinez-Fernandez A, Arrell DK, Lindor JZ, Dzeja PP, Ikeda Y, Perez-Terzic C, Terzic A (2011) Somatic oxidative bioenergetics transitions into pluripotency-dependent glycolysis to facilitate nuclear reprogramming. Cell Metab 14:264–271
14. Gogvadze V, Zhivotovsky B, Orrenius S (2010) The Warburg effect and mitochondrial stability in cancer cells. Mol Aspects Med 31:60–72
15. Diaz-Ruiz R, Rigoulet M, Devin A (2011) The Warburg and Crabtree effects: on the origin of cancer cell energy metabolism and of yeast glucose repression. Biochem Biophys Acta 1807:568–576
16. (2009) Nearly a century later, new findings support Warburg theory of cancer. Boston College. http://www.bc.edu
17. Pacini N, Borziani F (2014) Cancer stem cell theory and the Warburg effect, two sides of the same coin? Int J Mol Sci 15:8893–8930
18. Ferreira LM (2010) Cancer metabolism: the Warburg effect today. Exp Mol Pathol 89:372–380
19. Curing cancer: Nobel Laureate Otto Warburg. http://www.sparkpeople.com/mypage_public_journal_individual.asp?
20. Crabtree HG (1928) The carbohydrate metabolism of certain pathological overgrowths. Biochem J 22(5):1289–1298
21. Ibsen KH (1961) The Crabtree effect: a review. Cancer Res 21(7):829–841
22. Pfeiffer T, Morely A (2014) An evolutionary perspective on the Crabtree effect. Front Mol Biosci 1:17

23. Crabtree effect. http://www.bioblast.at/index.php/Crebtree effect. Accessed 31 Aug 2015
24. Efraim Racker. Wikipedia, the free encyclopedia
25. Racker E (1976) Why do tumor cells have a high aerobic glycolysis? J Cell Physiol 89:697–700
26. Vander Heiden MG, Locasale JW, Swanson KD, Sharfi H, Heffron GJ, Amador-Noguez D, Christofk HR, Wagner G, Rabinowitz JD, Asara JM, Cantley LC (2010) Evidence for an alternative glycolytic pathway in rapidly proliferating cells. Science 329:1492–1499
27. Zhang J, Khvorostov I, Hong JS, Oktay Y, Vergnes L, Nuebel E, Wahjudi PN, Setoguchi K, Wang G, Do A, Jung HJ, McCaffery JM, Kurland IJ, Reue K, Lee WN, Koehler CM, Teitell MA (2011) UCP2 regulates energy metabolism and differentiation potential of human pluripotent stem cells. EMBO J 30:4860–4873
28. Zhu S, Li W, Zhou H, Wei W, Ambasudhan R, Lin T, Kim J, Zhang K, Ding S (2010) Reprogramming of human primary somatic cells by OCT4 and chemical compounds. Cell Stem Cell 7:651–655
29. Ciavardelli D, Rossi C, Barcaroli D, Volpe S, Consalvo A, Zucchelli M, De Cola A, Scavo E, Carollo R et al (2014) Breast cancer stem cells rely on fermentative glycolysis and are sensitive to 2–deoxyglucose treatment. Cell Death Dis 5, e1336. doi:10.1038/cddis.2014.285
30. Kim TH, Suh DH, Kim M-K, Song YS (2014) Metformin against cancer stem cells through the modulation of energy metabolism: special consideration on ovarian cancer. Biomed Res Int 2014, 132702. doi:10.1155/2014/132702, 11 pages
31. Palorini R, Votta G, Baletrieri C, Monestiroli A, Olivieri S, Vento R, Chiaradona F (2014) Energy metabolism characterization of a novel cancer stem cell line 3–AB–OS. J Cell Biochem 115:368–379
32. Ito K, Suda T (2014) Metabolic requirements for the maintenance of self-renewing stem cells. Nat Rev Mol Cell Biol 15:243–256
33. Shen Y-A, Wang C-Y, Hsieh Y-T, Chen J-J, Wei Y-H (2015) Metabolic reprogramming orchestrates cancer stem cell properties in nasopharyngeal carcinoma. Cell Cycle 14(1):89–98
34. Pabst C, Krosl J, Fares I, Boucher G, Ruel R, Marinier A, Lemeieux S, Hebert J, Sauvageay G (2014) Identification of small molecules that support human leukemia stem cell activity ex vivo. Nat Methods 11:436–442. doi:10.1038/nmeth.2847
35. Forman NG, Wilson DF (1982) Energetics and stoichiometry of oxidative phosphorylation from NADH to cytochrome c in isolated rat liver mitochondria. J Biol Chem 257:12908–12915
36. Prigione A, Fauler B, Lurz R, Lehrach H, Adjaye J (2010) The senescence–related mitochondrial/oxidative stress pathway is repressed in human induced pluripotent stem cells. Stem Cells 28:721–733
37. Shultz M (2008) Rudolf Virchow. Emerg Infect Dis 14(9):1480–1481

Chapter 8
Metabolism in Cancer Stem Cell

Do not let yourself be tainted with a barren skepticism.

Louis Pasteur

Accumulating evidence suggests that the persistent activation of aerobic glycolysis in tumor cells plays a crucial role in cancer development; the inhibition of the increased glycolytic capacity of malignant cells may therefore represent a key anti-cancer strategy. The CSCs biology probably would not reach this level if there were not at least five scientists that have indicated and proved the essence of its particularities: Rudolf Virchow (1821–1902), Louis Pasteur (1822–1895), Otto Heinrich Warburg (1883–1970), Herbert Grace Crabtree (1890–XX), and Efraim Racker (1913–1991) [1–9].

Rudolf Virchow is famous for his quote: "*Omnis cellula e cellulae*" (All cells from one cell) which was so supportive for stem cell concept as well as for his detection of small cells that used to develop teratocarcinoma, which today are probably one of the versions of adult stem cells all over the body, isolated by M. Ratajczak, characterized and named VSELSCs [3] (Fig. 8.1).

Louis Pasteur made some of the greatest breakthroughs in modern medicine at the time, including pasteurization and vaccines for anthrax bacillus and rabies virus. He was also interested in metabolism of cancer cells. Cells can obtain energy through the oxygen-dependent pathway of oxidative phosphorylation (OXPHOS) and through the oxygen-independent pathway of glycolysis. Since OXPHOS is more efficient in generating ATP than glycolysis, it is recognized that the presence of oxygen results in the activation of OXPHOS and the inhibition of glycolysis (Pasteur effect). Thus, Pasteur effect is the inhibition of glycolysis by oxygen (aerobic conditions) due to inhibition of the enzyme Phosphofructokinase. Glycolytic intermediates from fructose 1,6-biphosphate onwards decrease while the earlier intermediates accumulate.

Otto Heinrich Warburg demonstrated in 1924 that cancer cells show an increased dependence on glycolysis to meet their energy needs, regardless of whether they were well oxygenated or not. Converting glucose to lactate, rather than metabolizing it through oxidative phosphorylation in the mitochondria, is far less efficient as less

© The Author(s) 2015
M. Pavlovic, B. Balint, *Bioengineering and Cancer Stem Cell Concept*,
DOI 10.1007/978-3-319-25670-2_8

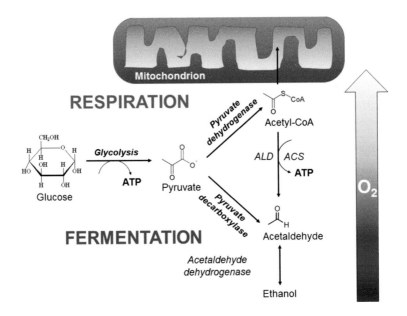

Fig. 8.1 Essential metabolic pathways in the cell

ATP is generated per unit of glucose metabolized [1, 4–8]. Therefore, a high rate of glucose uptake is required to meet increased energy needs to support rapid tumor progression [10–12]. He was awarded Nobel Prize in 1931 for his work. Previously, it was thought the Warburg effect was a consequence of damage to the mitochondria or an adaptation to hypoxic conditions during the early avascular phase of tumor development. Current insight revealed aerobic glycolysis supports various biosynthetic pathways and, consequently, the metabolic requirements for proliferation [13–19]. The PI3K pathway is considered to be a major determinant of the glycolytic phenotype through AKT1 and mTOR signaling, and subsequent downstream [20]. Hypoxia Inducible Factor 1 (HIF-1) transcription factor activation. Another important pathway is the AMP-activated protein kinase (AMPK) pathway. AMPK is often considered a metabolic checkpoint as it can control cell proliferation when activated under energetic stress, and activation of AMPK results in the inhibition of mTOR activity. Mutations in tumor suppressor genes such as LKB1, involved in AMPK activation, have been identified in certain cancers. Inhibitors of the glycolytic pathway, such as dichloroacetic acid (DCA) and 2-deoxy-D-glucose (2DG), are now being used in clinical studies as potential anticancer agents [20].

According to Warburg, damaged cell respiration causes fermentation, resulting in acidity (low pH) at the cellular level. Since fermentation was a main metabolic pathway in cancer cell, Warburg reported that cancer cells maintain a lower pH as low as 6.0 due to presence of lactic acid and elevated CO_2. He essentially proved that cancer cannot grow or develop in body alkalinity of 7.46. Once in the imbalanced pH conditions, cancer cannot survive an alkaline cellular pH level, nor the presence of highly oxygenated cells. Thus, according to him, low cellular oxygen is a primarily causal factor of cancer [12]. In a lack of oxygen, cells revert to primary nutritional program to nourish itself by converting glucose through the process of fermentation. The by-product of fermentation, lactic acid, lowers the cell pH causing acid/alkaline disbalance and destroys the ability of DNA and RNA to control cell division [13] (Fig. 8.2).

Named after the English biochemist *Herbert Grace Crabtree*, the Crabtree effect (1929) describes the phenomenon whereby the yeast, *Saccharomyces cerevisiae*, produces ethanol (alcohol) in aerobic conditions and high external glucose concentrations rather than producing biomass via the tricarboxylic acid (TCA) cycle, and in that way suppresses respiration by that [10–12]. Later on, it has been shown that the same effect does exist in tumor cells including CSCs [17–21]. The main differences between Warburg and Crabtree effects are that Crabtree effect occurs as a short-term adaptive mechanism after glucose-induced repression, while Warburg effect is a long-term adaptive mechanism [22].

Carcinogenesis occurs through a series of steps from normal into benign and finally malignant phenotype. This cancer evolutionary trajectory has been accompanied by similar metabolic transformation from normal metabolism into Pasteur and/or Crabtree Effects into Warburg Effect and finally Cannibalism and/or Lactate-Symbiosis [23]. Due to lactate production as an end product of glycolysis, tumor colonies acquire new phenotypes that rely on lactate as energetic fuel. Presence of Warburg Effect indicates that some tumor cells undergo partial (if not complete) de-endosymbiosis and so cancer cells have been become unicellular microorganism (anti-Dollo's Law) specially when they evolve to develop cannibalism as way of metabolism while oxidative types of cells that rely on lactate, as their energetic fuel,

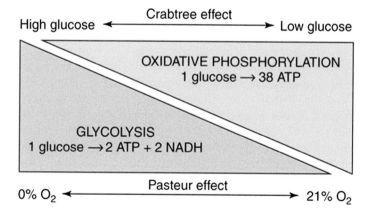

Fig. 8.2 The relationship between Crabtree and Pasteur effect in cancer cell

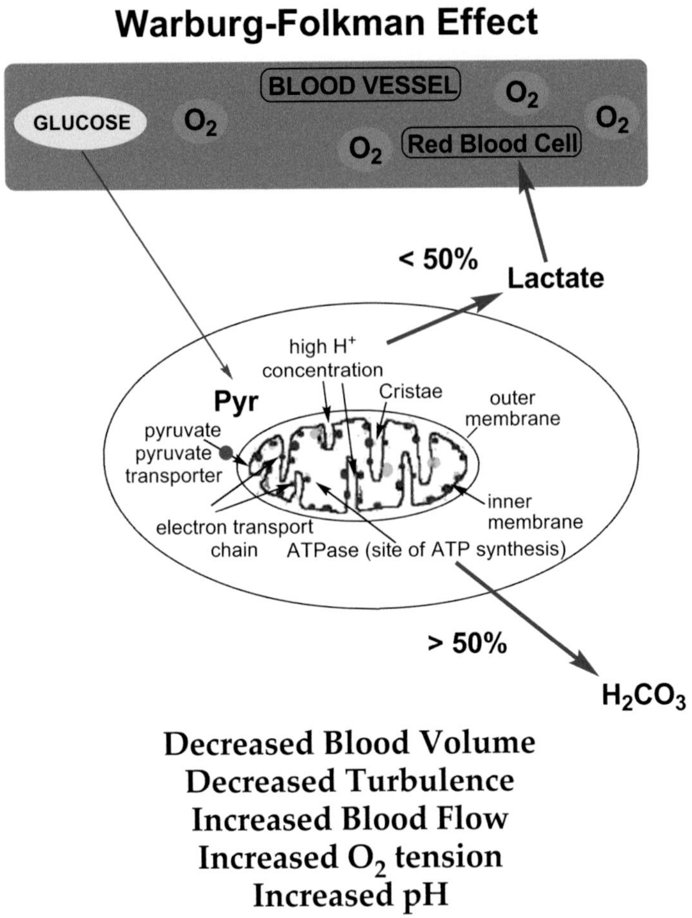

Warburg-Folkman Effect

Fig. 8.3 Warburg's effect (known also as Warburg-Folkman effect) in cancer cell and significance of pH in tumor

might represent extra-endosymbiosis [23]. Thus, at the end, the cancer colony could be considered as integrated metabolic ecosystem [23] (Fig. 8.3).

Proper understanding of tumor metabolism will contribute to discover potential anticancer agents besides conventional chemotherapy. A great contribution to this field was already given by *Efraim Racker* (1913–1991) who was working on demonstration of dependence of glycolysis in ATP-ase and the continuous regeneration of ADP and phosphate. In 1960, with his group he has discovered the F1 subunit of the first enzyme of oxidative phosphorylation, a necessary coupling factor for ATP-ase. Parallel with that contribution he was able to prove the Peter Mitchel's hypothesis that contrary to popular opinion, ATP synthesis was not coupled to respiration through a high-energy intermediate but instead by a transmembrane proton gradient (proton-motive force of the respiratory chain) [12, 13, 22].

This field is current topic, giving the new light and future directions based upon the knowledge that has been accumulated for a long time. The metabolic signatures of CSCs and the underlying mechanisms of their stem cell-like properties remain still elusive. It seems that there is still a lot of work to be done on each particular CSC. Altogether, accumulated data indicate that metabolic reprogramming is critical for CSCs to sustain self-renewal, escape from differentiation, and enhance the antioxidative defense mechanism [22]. To reach each characteristic, metabolic reprogramming would probably be the goal for successful design of novel therapeutic strategies through metabolic interventions in CSCs.

References

1. Pacini N, Borziani F (2014) Cancer stem cell theory and the Warburg effect, two sides of the same coin? Int J Mol Sci 15:8893–8930. doi:10.3390/ijms15058893
2. Yuan S, Wang F, Chen G, Zhang H, Feng L, Wang L, Colman H, Keating MJ, Li X, Xu RH, Wang J, Huang P (2013) Effective elimination of cancer stem cells by a novel drug combination strategy. Stem Cells 31(1):23–34. doi:10.1002/stem.1273
3. Shultz M (2008) Rudolf Virchow. Emerg Infect Dis 14(9):1480–1481
4. Gogvadze V, Zhivotovsky B, Orrenius S (2010) The Warburg effect and mitochondrial stability in cancer cells. Mol Aspects Med 31:60–72
5. Diaz-Ruiz R, Rigoulet M, Devin A (2011) The Warburg and Crabtree Effects: on the origin of cancer cell energy metabolism and of yeast glucose repression. Biochim Biophys Acta 1807:568–576
6. Nearly a century later, new findings support Warburg Theory of cancer. (2009) Boston College. http://www.bc.edu
7. Ferreira LMR (2010) Cancer metabolism: the Warburg effect today. Exp Mol Pathol 89:372–380
8. Curing cancer: Nobel Laureate Otto Warburg. http://www.sparkpeople.com/mypage_public_journal_individual.asp?
9. Crabtree HG (1928) The carbohydrate metabolism of certain pathological overgrowths. Biochem J 22(5):1289–1298
10. Pfeiffer T, Morely A (2014) An evolutionary perspective on the Crabtree effect. Front Mol Biosci 1:17
11. Crabtree effect. http://www.bioblast.at/index.php/Crabtree_effect. Accessed 31 Aug 2015
12. Efraim Racker. Wikipedia, the free encyclopedia
13. Racker E (1976) Why do tumor cells have a high aerobic glycolysis? J Cell Physiol 89:697–700
14. Vander Heiden MG, Locasale JW, Swanson KD, Sharfi H, Heffron GJ, Amador-Noguez D, Christofk HR, Wagner G, Rabinowitz JD, Asara JM, Cantley LC (2010) Evidence for an alternative glycolytic pathway in rapidly proliferating cells. Science 329:1492–1499
15. Zhang J, Khvorostov I, Hong JS, Oktay Y, Vergnes L, Nuebel E, Wahjudi PN, Setoguchi K, Wang G, Do A, Jung H-J, McCaffery JM, Kurland IJ, Reue K, Lee W-NP, Koehler CM, Teitell MA (2011) UCP2 regulates energy metabolism and differentiation potential of human pluripotent stem cells. EMBO J 30:4860–4873
16. Zhu S, Li W, Zhou H, Wei W, Ambasudhan R, Lin T, Kim J, Zhang K, Ding S (2010) Reprogramming of human primary somatic cells by OCT4 and chemical compounds. Cell Stem Cell 7:651–655

17. Ciavardelli D, Rossi C, Barcaroli D, Volpe S, Consalvo A, Zucchelli M, De Cola A, Scavo E, Carollo R et al (2014) Breast cancer stem cells rely on fermentative glycolysis and are sensitive to 2-deoxyglucose treatment. Cell Death Dis 5, e1336. doi:10.1038/cddis.2014.285
18. Kim TH, Suh DH, Kim M-K, Song YS (2014) Metformin against cancer stem cells through the modulation of energy metabolism: special consideration on ovarian cancer. Biomed Res Int 2014:132702. doi:10.1155/2014/132702
19. Palorini R, Votta G, Baletrieri C, Monestiroli A, Olivieri S, Vento R, Chiaradona F (2014) Energy metabolism characterization of a novel cancer stem cell line 3-AB-OS. J Cell Biochem 115:368–379
20. Ito K, Suda T (2014) Metabolic requirements for the maintenance of self-renewing stem cells. Nat Rev Mol Cell Biol 15:243–256
21. Shen Y-A, Wang C-Y, Hsieh Y-T, Chen J-J, Wei Y-H (2015) Metabolic reprograming orchestrates cancer stem cell properties in nasopharyngeal carcinoma. Cell Cycle 14(1):89–98
22. Song CW, Griffin R, Park HJ (2006) Influence of tumor pH on Therapeutic response. In: Teicher B (ed) Cancer drug discovery and development: cancer drug resistance. Humana Press Inc., Totowa, pp 21–40
23. Alfarouk KO, Shayoub MEA, Muddathir AK, Elhassan GO, Bashil A (2011) Evolution of tumor metabolism might reflect carcinogenesis as a reverse evolution process (Dismantling of Multicellularity). Cancer 3:3002–3017

Chapter 9
Different Approaches for Anticancer/ Antitumor Therapy

Extraordinary claims require extraordinary evidence.

Carl Sagan

Introduction

Science and its community are still being challenged by the search of the best cancer treatment method, which could put end to this stubborn disease. Statistics show more than a million men and women are diagnosed with cancer and practically half of these patients died due to this disease in 2012 only in the USA [1]. Different treatments vary and depend on factors mainly showed by the patient status. At some point in the past, surgery and radiation therapy were the most preferred methods of treatments when the disease was localized [2, 3]. Nonetheless, both could not sufficiently control the metastasis of cancer [4]. Consequently, different treatment methods, mostly known as cancer-targeted therapies and drug delivery development, are currently applied.

This field is broad and involves targeted therapy such as in cancer stem cells (CSCs), heat shock proteins, and multiple signaling pathways [1–9]. Additionally, as part of these methods, therapy can be accomplished by the manipulation of antibodies, small molecules, and ligands. These targeted therapy methods are basically new in the research field and have a great potential to finally develop into the best cancer treatment available.

Cancer-Targeted Therapy

There are no successful improvements in mortality rates for patients with metastatic cancer for a long time since the 1970s, which simply confirms that cancer is still one of the most severe diseases. Nevertheless, research has helped improve the

M. Pavlovic, B. Balint, *Bioengineering and Cancer Stem Cell Concept*, DOI 10.1007/978-3-319-25670-2_9 knowledge in particular and specific tumor entities and therapies [10–17]. Targeted therapy may directly target tumor antigens by altering their signaling or indirectly depending on these tumor antigens being expressed on the cell surface, which behave as aims for effector molecules-containing ligands [4].

Antibody-Targeted Therapy

Monoclonal antibodies (MoAbs), which are approved by the US FDA due to their effectiveness for different cancer treatments, have the ability to target tumors improving the selectivity of other anticancer agents [17–24] (Table 9.1).

Cancer therapy with antibodies may be efficient alone as well as in combination with chemotherapy and radiation therapy [24–28]. It is important that the antigen must be reactive with the target cell and not the healthy tissue. Thus, if antigens are not found on the cell surfaces but instead circulating in the peripheral blood, they should be prevented from reaching cancer cells.

Murine, rabbit, or rabbit proteins were the first antibodies studied and purified [4, 15, 25]. In humans, these antibodies do not normally promote antibody-dependent cell-mediated cytotoxicity (ADCC) and complement-dependent cytotoxicity (CDC) which would assist in the destruction of malignant cells. Thus, humanized antibodies or hybrids are engineered allowing the human antibodies to be linked with a murine backbone, for instance, making a chimera of antibodies from two different species, which increases the effectiveness of therapy. These chimeric antibodies have been able to activate immune effector's functions such as ADCC and CDC becoming a great success in MoAb-based therapies in the medical field [4]. The list of these targeted molecules is very long [24–44].

Small Molecules-Targeted Therapy

This type of targeted therapy involves chemotherapy with the purpose of supplying agents specifically for cancer cells. For instance, lead compounds may be used as inhibitors for molecular targets [4]. Additionally, protein kinases have been targeted the most by small molecules inhibitors because they do not only regulate cell life such as protein phosphorylation but may be related to pathways in tumors due to abnormal phosphorylation. Consequently, drug development increased specifically for targeting membrane associated protein tyrosine kinases [4]. Some examples are: Naphthalene sulfonamides, which work as calcium-binding protein calmodulin antagonist, Glivec, which is the first selective inhibitor of tyrosine kinase for cancer treatment, and Gefitinib, which is a selective inhibitor for oral epidermal growth factor receptor tyrosine kinase [4, 45] (Fig. 9.1).

Table 9.1 FDA approved therapeutic antibodies

Monoclonal antibody	Target	Indication	Product	Year	Corporate sponsors	References
Rituximab (Rituxan)	CD20	Low-grade B-cell NHL	Chimeric	1997	IDEC, Genentech	[21–24]
Trastuzumab (Herceptin)	HER2/neu	Metastatic breast cancer	Humanized	1998	Genentech	[17, 19, 25]
Gemtuzumab-ozogamicin (Mylotarg)	CD33	Acute myeloid leukemia	Humanized	2000	Wyeth Laboratories	[26–28]
Alemtuzumab (Campath)	CD52	Chronic lymphocytic leukemia	Humanized	2001	Millennium and ILEX Partners	[29–31]
Ibritumomab-tiuxetan-^{90}Y (Zevalin)	CD20	Non-Hodgkin lymphoma	Mouse	2002	IDEC	[32–34]
Tositumomab/Tositumomab-^{131}I (Bexxar)	CD20	Non-Hodgkin lymphoma	Mouse	2003	Corixa, GlaxoSmithKline	[35–37]
Cetuximab (Erbitux)	EGFR	Metastatic CRC; HNSCC	Chimeric	2004	ImClone Systems	[37, 38]
Bevacizumab (Avastin)	VEGF	Metastatic CRC; NSCLC	Humanized	2004	Genentech	[41–44]

NHL non-Hodgkin lymphoma, *CRC* colorectal cancer, *HNSCC* head and neck squamous cell carcinoma, *NSCLC* non-small cell lung cancer

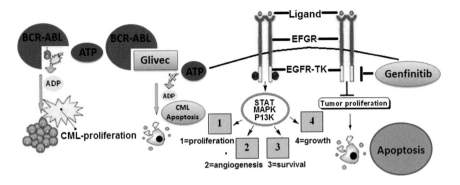

Fig. 9.1 Mechanisms of small molecule inhibitors. (**a**) Constitutively activated Bcr-Abl tyrosine kinase causes chronic myelogenous leukemia (CML). (**b**) EGFR-mediated signaling contributes to the up-regulation of essential processes for tumor growth and progression

Ligand-Targeted Therapy

The efforts to increase levels of toxicity in cancer cells with anticancer chemotherapeutics are limited because molecular targets may not differentiate/distinct well enough cancer cells from normal cells due to their common features. Normal tissues such as bone marrow or gastrointestinal tract may be affected inducing side effects in a patient. However, ligand-targeted therapy helps develop specific tumor detection and reduces toxicity levels. Even though this therapy shows potential favorable results, there are certain obstacles to overcome. Human tumors actually show genetic diversity yielding drug-resistant cells [45]. These cells have resistant mechanisms, which affect targeted drugs and antitumor agents by mutating cell death pathways, for instance. Another case is high tumor interstitial fluid pressure (IFP), which decreases the transcapillary transport of drugs or therapeutic antibodies in tumors consequently increasing toxicity. Regardless of this barrier for drug development, increased tumor IFP's mechanisms are still not completely understood but possibly involving blood vessel leakiness, lymph vessel abnormalities, interstitial fibrosis, and contraction of the interstitial space mediated by stromal fibroblasts [4]. Even though high IFP is correlated with high recurrence rate and poor prognosis for patients with cervical cancer receiving radiation therapy, ligand therapy may increase the buildup of drugs in high IFP of the tumor by using the ligand affinity to the plasma membrane receptors in cancer cells.

Thus far, a technique called phage display is used for the search of targeting ligands [4]. In this case, a peptide is expressed as fusion with a coat of protein of bacteriophage. When the targeting ligands to cancer cells are found, they are conjugated with chemotherapeutic drugs in order to kill cancer cells. Liposomes are chosen to conjugate with targeting ligands because of prolonged blood circulation, sufficient tumor accumulation, and controlled drug release. This targeted liposomal system may be definitely improved by ligand-targeted therapy. One positive result is that targeting liposomes have an enhanced antitumor effect [4].

Finally, another important factor that could make this therapy not efficient enough is the natural properties of CSCs, which may increase the resistance to chemotherapy agents. As a consequence, relapse could occur by insufficient targeting of the CSCs during initial treatment. Tumor pathogenesis is linked to these characteristics of stem cells, which are found in various types of cancer. However, understanding efficiently the nature of stem cells will help improve the detection of premalignant lesions and tumors through different targeted therapies [4, 45–55] (Fig. 9.2).

Isocitrate dehydrogenase type 1 (IDH1) is represented as monoallelic point mutations in the development of gliomas, which are glial tumors, developed from glial cells, and other tumor types [7, 8]. IDH1 also represents a possible target for immunotherapy since it is a tumor-specific neoantigen in all tumor cells containing an immunogenic epitope appropriate for mutation-specific vaccination. Since IDH1 is present in all tumor cells of gliomas, a mutation-specific anti-IDH1 vaccine may induce antitumor immunity becoming another therapeutic strategy for these types of tumors [7].

Kinase Inhibitors Targeting Multiple Signaling Pathways

There are many cellular signaling pathways, which are able to promote cancer development. Some examples are the phosphatidyl inositol-3-kinase (PI3K)/AKT, protein kinase C (PKC) family, and mitogen-activated protein kinase (MAPK)/Ras

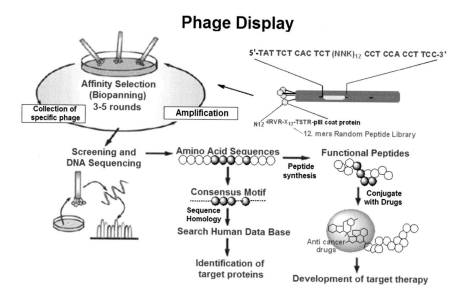

Fig. 9.2 Identification of targeting ligands to cancer cells by phage display; vaccine targeting mutant

signaling cascades. The main characteristic among them is that the receptor kinases (RTKs) are part of the signal transduction in cancer cells. Inhibition of kinases has shown therapeutic benefit associated with molecular targeted anticancer agents. An important factor in targeted anticancer therapy is the understanding of these signaling pathways, which promote malignant cell transformation and proliferation [5]. As part of this process, there is a greater emphasis on targeting multiple signaling pathways by an individual agent because cancer produces multiple abnormalities instead of a single defect. Growth receptors and their signal transduction and tumor angiogenesis (process by which new vessels grow from pre-existing vessels in the surrounding normal tissues) [8] are key therapy targets, particularly the latter because it may spare normal cells during therapy [5] (Fig. 9.3).

Attacking Multiple Targets

Multitargeted tyrosine kinases inhibitors target RTKs on tumor cells as well as supporting cells such as vascular endothelial growth factor receptors (VEGFRs) and platelet-derived growth factor receptors (PDGFRs) which indicate the antitumor effects and anti-angiogenic properties. Studies emphasize the effectiveness of anticancer therapies involving different drug combinations. There are benefits when targeted anti-angiogenic agents are combined with conventional cytotoxic therapies by enhancing anticancer effects of anti-angiogenic treatment. Additionally, targeted therapies in combination with traditional chemotherapy may overcome resistance of breast cancer (resistance to antiestrogen therapy) specifically [5, 25] (Table 9.2).

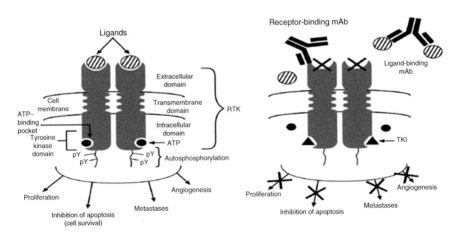

Fig. 9.3 Structure of RTKs and mechanism of action of monoclonal antibodies and TKIs inhibiting the shown receptors. pY, phosphorylated tyrosine

	Membrane tyrosine kinases	Intracellular kinases
Table 9.2 Major kinases proposed as molecular targets and expressed on cell membrane and cytoplasm	EGFR	Ras
	HER2	Raf
	PDGFR	MAP
	FGFR	MEK
	IGF-1R	mTOR
	VEGFR	PKC
	FLT3	AURORA TK
	RET	BCR-ABL

Agents Targeting Serine/Threonine Kinases

Serine and Threonine kinases are involved in growth, proapoptotic signaling, and angiogenesis. The PI3K/AKT pathway is involved in cancers via regulatory effects on mammalian cell proliferation and survival. PI3K is composed of the p85 regulatory subunit and a p110 catalytic subunit. When PI3Ks are activated, they catalyze the phosphorylation of phosphatidylinositol trisphosphate (PIP$_3$), which serves as a ligand to recruit AKT to the plasma membrane. AKT is phosphorylated by serine/threonine kinase making AKT active in order to promote cell proliferation, growth, and survival of cancer cells. Mammalian target of rapamycin (mTOR) is a crucial AKT substrate in charge of cellular metabolism, signaling cells to expand, grow and proliferate, and may regulate apoptotic processes. It is also suggested that the abnormality of these processes promotes cancer development and, therefore, mTOR inhibitors are assessed to become anticancer agents.

Another signaling pathway being targeted involves the protein kinase C (PKC) family, which is important transducer in the production of diacylglycerol (DAG) (intracellular second messenger). PKC inhibitors are possible anticancer agents since PKCs may have an active participation in carcinogenesis and malignant transformation (Fig. 9.4).

Also, there is another type of serine/threonine kinase called Raf kinase, an effector of Ras in the MAPL/Ras signal transduction pathway, which is linked to cancer. The MEK/ERK cascade is often observed in human cancers. Raf kinase inhibition by sorafenib (oral multikinase inhibitor) interrupts this cascade, inhibiting both the kinase activity correlated to various RTKs and the MAPK/Ras pathway associated with RTK activation. Even though sorafenib is a multitargeted TKI in late stages of clinical development, it has already shown inhibition activity in the ERK pathway phosphorylation for cultured breast, melanoma, pancreatic, and colon cancer cell lines. It shows antitumor activity in many xenograft models without increasing toxicity when combined with anticancer drugs [5, 45].

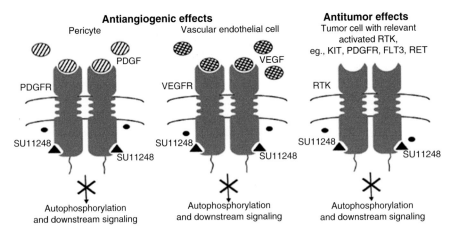

Fig. 9.4 Mechanism of action of anti-angiogenic and antitumor effects by SU11248

Heat Shock Protein Targets

Another type of therapy used in cancer treatment aims for heat shock proteins (HSPs), which are synthesized by cells exposed to some sort of stress in order to protect the latter from damage. They can be found overexpressed in many cancers by promoting tumor cell proliferation and inhibiting cellular death pathways. Also known as stress proteins, HSPs function as molecular chaperones, which compose from 5 to 10 % of total protein in a healthy cell. Nevertheless, when the cell is under stress, proteins misfold or accumulate and bind to chaperones in order to release a heat shock factor (HSF), which promotes the cellular increase of HSP concentration. As a result, HSPs facilitate the refolding of damaged proteins or seize them for degradation.

In cancer, the overexpression of HSPs is clear and described as an adaptive response by cancer cells to keep protein homeostasis, which leads to the survival of the cell in a stressed environment. There are different types of HSPs mainly classified based on their molecular size describing the larger HSPs (HSP100, HSP90, HSP70, HSP60, HSP40) as ATP-dependent while the smaller HSPs (sHSPs) as ATP-independent. Chaperones, such as HSP90, may be highly expressed in most tumor cells.

Overall, HSPs may contribute in six cell physiology alterations in order to identify cancerous growth. One alteration is described as *self-sufficiency in generating growth signals in cancer cells* where HSP90, for instance, keeps the activities of proto-oncogenes, such as HER2 proto-oncogene, and protein kinases, such as Akt, c-SRC and Raf-1, in order to enhance tumor growth and survival. Additionally, it promotes the stabilization of mutant protein conformations resulting in the accumulation of mutated molecules in cancer cells [3]. The second alteration is *insensitivity to antiproliferative signals where HSP70 is the main participant by binding to p53 tumor suppressor protein.* Even though the mutation in p53 protein provokes cancer development, there is still no evidence that increased levels of HSP70 may mutate p53 inducing malignant growth. The third alteration is the *avoidance of apoptosis where HSPs play an extensive role in the programmed cell death.* When HSP70 or HSP27 is inactivated, caspase-dependent apoptosis, for instance, occurs while the up-regulated expression of these chaperones inhibits the apoptosis process [3]. The fourth alteration is the *unlimited replicative capacity*, which allows cancer cells to escape senescence by avoiding the shortening of telomeres. HSP90 helps stabilize telomeres while HSP75 (HSP70 family) plays a role in increasing cancer cell divisions and counter senescence by inhibiting p53 activity. The fifth alteration is *tumor angiogenesis*, which is mainly enhanced by the overexpression of HSP90. The sixth alteration is *invasive and metastatic capability of tumors*, which, according to studies, correlates with the increase of HSP27 and HSP70. HSP90 also leads to the increased cancerous invasion by cleaving constituents of the extracellular matrix.

Due to the diverse roles of HSPs, they become easy targets for cancer therapy. HSP inhibitors may actually obstruct multiple signaling pathways, which lead to the specific apoptosis of cancer cells [3].

Cancer Stem Cells Therapy: Concept and Novelties

As previously stated, CSC therapy could improve the effectiveness in cancer treatment by targeting these particular cells. CSCs are characterized as self-renewal, heterogeneous, and apoptotic resistant, which may contribute to the decrease of successful conventional therapies on differentiating tumor cells. CSCs arise from stem cells (SCs) or precursor cells when the latter are originally normal and transform into CSCs due to the accumulation of genetic modifications and epigenetic alterations such as the deregulation of signaling pathways (Fig. 9.5).

When SCs divide symmetrically, depending on the organism's needs, the daughter cells either differentiate or remain undifferentiated. Differentiation is normally induced by the sensitivity of the niche signaling sensitivity and outer cell environment. When SCs are cultured in vitro, they proliferate and differentiate rapidly. In order to keep them undifferentiated, they are cultured under special conditions so they can be stimulated to produce new cells. However, it also suggests that CSCs may be affected and start promoting the initiation of tumor growth. Interestingly enough, CSCs do not produce cancer when displaced into a different niche where epigenetic changes are not present while SCs initiate tumor growth when placed in a tissue damaged by radiation, for instance. Additionally, CSCs may be formed due to alterations in signal transmission and genetic factors [6, 55–65].

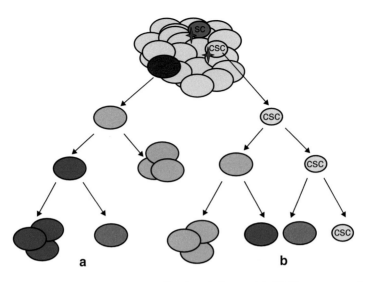

Fig. 9.5 Tumor development models. (**a**) Clonal evolution model. (**b**) Cancer stem cell model. *Green*=niche cells. *Blue*=stem cell. *Yellow*=cancer stem cell. *Red* star=adhesive molecules. *Brown, orange, red* and *dark* turquoise=cells with genetic alterations accumulation

Possibilities of Engineering Targeted Cancer SC Therapy Using Principles of Magnetism

The fundamental goal of cancer therapy lies in a few key ideas: (1) create as little side effect of the treatment to the host's tissues, (2) treat as noninvasively as possible, and (3) have long-term viability of treatment as stem cells vary in their genotypic expression. The third point, however, may be debatable if the CSCs can be obliterated in the first treatment.

As technological progression builds new opportunity in other realms of science, so does it in the world of medicine. Two new treatment modalities are on the forefront of oncological intervention: *alternating magnetic fields* on replicating CSCs. Although in the early phases of testing, it shows promises of accomplishing limited-to-no side effects as well as being as noninvasive as possible. This novel treatment could be used in combination with chemotherapy or radiotherapy, as indicated [45].

The other novel therapy targeting CSCs that has shown great result is the use of alternating magnetic fields. As dividing cells undergo the various stages of cell replication, *a developmental stage known as mitosis is the target of this therapy.* During mitosis, all of the sister chromatids are lined up along the midline of the cell and still adjoined to one another by a centromere, which then become the target of spindles emergent from the centrioles at opposite poles of the dividing cell [45]. *These spindles have a polarity in charge due to their molecular composition*

(Figs. 9.6 and 9.7). If this transient treatment field (TTF) is applied via an external array, the spindles of CSCs (which are smaller than other progenitors and normal cells) are disrupted by the alternating fields and a resultant disruption of cancer cell replication is accomplished [38]. It is important to note, however, that CSCs are smaller than typical, normal-state mitotic cells and can therefore be targeted with specific frequencies in order to minimalize damage to healthy, non-cancerous cells (Fig. 9.7).

The first clinical trial was in 2003 for patients with glioblastoma (GBM), the most aggressive and most common form of primary brain tumor in the United States. Two years later, three of the original ten patients were still alive, two of which had no progression of the cancer whatsoever. In 2011, the FDA approved TTF as a viable treatment for GBM [45].

Currently, clinical trials are being run for the utilization of TTF with lung cancer, as well as in vitro research for many other types of cancer, including cervical. Some devices already are produced and utilized in the market, making the treatment more

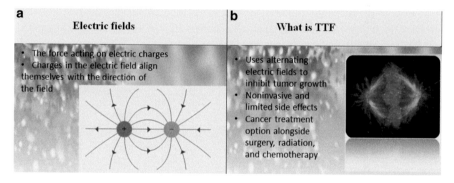

Fig. 9.6 (**a**) Electric fields (**b**) Stem-line therapy

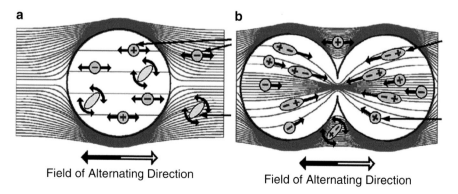

Fig. 9.7 Fields of alternating directions with polarization and difference in size of normal and cancer cells

readily available. NovocureTM, a commercial stage private oncology company, manufactures the device, NovoTTF™-100A, a wearable device weighing around 6 lb that can fit into a shoulder bag for easy handling. Using noninvasive, insulated transducer arrays that are placed directly on the skin in the region surrounding the tumor, TTF therapy is unlike previous applications of electricity in medicine [45].

Increased apoptosis of CSCs will result in a significantly smaller number of matured cancer cells, which can then be addressed successfully with common anticancer therapies. Therefore, anticancer therapy that only results in apoptosis of the matured cancer cells and/or only inhibits the proliferation of CSCs provides a potential window of opportunity for new and more aggressive CSC mutants to occur and might be unsuccessful, if not dangerous. It is expected that the elimination of cancer should target the CSC pool, and successful treatment regimens would need to be the result of an orchestrated "target and destroy" effect. TTF therapy is a locally or regionally delivered treatment that uses electric fields within the human body that disrupt the rapid cell division exhibited by cancer cells. TTF therapy was developed to provide physicians and patients with a fourth treatment option for cancer in addition to surgery, radiation therapy, and chemotherapy (Figs. 9.8 and 9.9).

In light of the difficulty of the manipulation of the CSC model, the research that has been done thus far is providing a solid framework upon which a new, improved paradigm of oncological treatment will be:

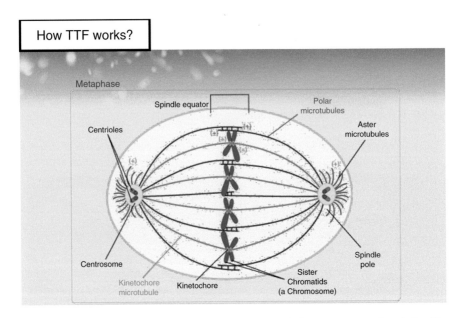

Fig. 9.8 Transient treatment field (TTF) inducing mitotic spindle rearrangement in polarity. The induced magnetic field specific to the frequency of CSCs disrupts the spindle formation and subsequent continuance of mitosis

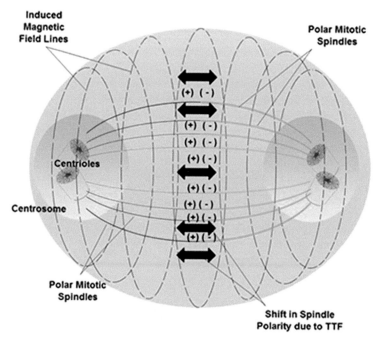

Fig. 9.9 Transient treatment field (TTF) inducing mitotic spindle rearrangement in polarity. The induced magnetic field specific to the frequency of CSCs disrupts the spindle formation and subsequent continuance of mitosis

- TTF therapy affects only one cell type at a time, and not cells that are in division.
- TTF therapy is not expected to affect the normal functions of bone marrow in creating red and white blood cells, since the bone marrow is naturally shielded from the fields.
- TTF therapy is delivered locally through a physical, nonchemical pathway, known as blood brain barrier (BBB). This allows the brain tumor treatment, whereas other mitotic inhibitor treatments such as taxanes and vinca alkaloids have poor diffusion across (BBB) and, therefore, are rarely used to treat brain tumors.
- There is no evidence of cumulative damage to healthy tissues in the body when exposed to TTF therapy. Since the fields alternate so rapidly, they have no effect on normal quiescent cells nor do they stimulate nerves and muscles [45].

Taken together, these properties will potentially allow patients to receive TTF treatment for as long as necessary with minimal side effects while maintaining a high quality of life established.

Cancer Therapy Outlook

The existence of CSCs was confirmed during the 1960s when cells were taken from specific sites in patients with malignancies in order to be transplanted to other parts resulting in that only a minor percentage of transplanted cells produced a tumor (animal models were later used due to ethical controversies).

CSCs may be resistant to therapy due to the increased expression of proteins from the BCL-2 family since they lead to the increase of expression of membrane proteins accountable for drug resistance. There are reports that checkpoint proteins play a critical role in determining the resistance of CSC to radiotherapy. These checkpoint proteins are activated and overexpressed due to DNA damaged [6] (Table 9.3).

There are two models of cancer growth, which explain the development of tumors. The *stochastic model* states that every cancer cell is capable of proliferating and regenerating a tumor. The other model is, of course, *the CSC model*, which, as stated before, states that only a small group of cells within the tumor population is capable of initiating and continuing tumor growth [45]. This suggests that knowledge of the cell surface phenotypes for hematopoietic stem cells (HSCs—blood cells derived from mesoderm and located in the red bone marrow) and progenitor cells may allow the isolation of similar subpopulations from different tumor types such as acute myeloid leukemia [2].

Chemoresistance, Radioresistance, and Reactive Oxygen Species as Targets

CSCs are most likely responsible for tumor chemoresistance and recurrence because of their unlimited self-renewal capacity. During experiments, in animal models, the multidrug resistance transporter breast cancer resistant protein 1 (BCRP1) was identified to be the molecular determinant of enriching CSCs. Additionally,

Table 9.3 Cell surface markers correlated with cancer stem cells

Tumor type	Cell surface markers	References
Acute myeloid leukemia	CD34+CD38−	[46]
Breast tumor	CD44+CD24−ESA+	[47]
Brain tumor	CD133+	[9]
Colon cancer	CD133+	[50, 53]
Head and neck cancer	CD44+	[52]
Prostate cancer	CD44+	[48, 51]
Metastatic melanoma	CD20+	[62]
Colorectal cancer	EpCAM^{high}CD44+CD166+	[63]
Pancreatic cancer	CD24+CD44+ESA+	[4]
Lung adenocarcinoma	Sca1+CD45−Pecam−CD34+	[64]
Bone sarcoma	Stro1+CD105+CD44+	[65]

neuroblastoma cells were more resistant to the effects of drugs and are contributors of the overall drug resistance phenotype of relapsed or resistant cancers besides sharing characteristics of tumor stem cells. Brain tumor stem cells increased resistance in response to treatment with chemotherapeutic agents. Clinically speaking, tumors respond to chemotherapies only to come back more aggressive because if the chemotherapy destroys most of the cells in a tumor, CSCs could be left behind in order to reoccur. Inhibitors of certain signaling pathways (Notch signaling pathway, for instance) show reduction of specific tumor stem cells such as in the brain, indicating a loss of tumor-forming capacity and finding a new therapeutic target.

CSCs also contributed to radiation resistance such as in highly malignant gliomas through activation of DNA damage checkpoint response and an increase in DNA repair. Specific inhibitors are necessary to reverse this radioresistance providing a therapeutic advantage to reducing brain tumor incidence, for instance. Since radiation treatment fails sooner or later because it may not kill subpopulation of tumor-initiating cells, creating therapies, which promote the differentiation of cancer cells, is the new goal.

Reactive oxygen species (ROS), such as superoxide (O^-_2) and its simultaneous oxidation and reduction (redox) product such as hydrogen peroxide (H_2O_2) have also been involved in the progression of cancer. Oxidative stress regulates various cellular processes correlated with transformed phenotype, including DNA damage, proliferation, cell adhesion and migration, and cell survival or death signaling. Therefore, in order to increase sensitivity to chemotherapeutic drugs and modulate tumor cell apoptosis, manipulation of cellular redox state is necessary. On top of that, there is a possibility that CSCs in malignant brain tumors, for instance, may have an altered ROS regulatory mechanism, which allows them to keep on self-renewing in response to chemotherapeutic agents [2] (Fig. 9.10).

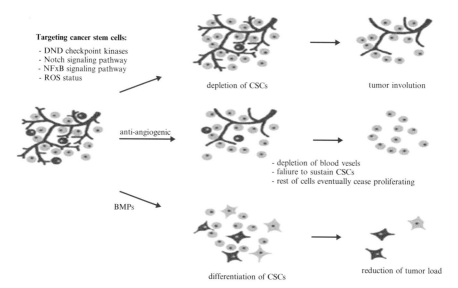

Fig. 9.10 Therapeutic strategies. (*Top*) Targeting of cancer stem cells and their niche. (*Middle*) Anti-angiogenesis. (*Bottom*) Bone morphogenetic proteins (BMPs)

Conclusions

Cancer research is still growing and their treatments becoming more effective than years ago. Nevertheless, more knowledge and understanding about this serious disease is necessary. Fortunately, the current efforts of taking different approaches besides surgery and radiation are being emphasized having diverse studies mainly focusing on cancer-targeted therapy [45–65]. Targeted therapies are changing the manner the scientific community treat cancer in order to design personalized treatments. Among the treatments that target specific agents at the molecular level, CSCs research seems to be the field where improvements may continue to be made because of their potential to develop into other diverse cell types, leading to get to the root of the problem by either eradicating or differentiating them. These targeted therapies definitely receive much attention over the past years and even though further studies need to be done, they could successfully treat or cure cancer.

References

1. Mavroudi M, Zarogoulidis P, Porpodis K, Kioumis I, Lampaki S, Yarmus L, Malecki R, Zarogoulidis K, Malecki M (2014) Stem cell's guided therapy of cancer: new frontier in personalized and targeted therapy. J Cancer Res Ther (Manch) 2(1):22–23
2. Tang C, Ang BT, Pervaiz S (2007) Cancer stem cell: target for anti-cancer therapy. FASEB J 21:3777–3785
3. Soo ETL, Yip GWC, Lwin ZM, Kumar SD, Bay B (2008) Heat shock proteins as novel therapeutic targets in cancer. In Vivo 22:311–316
4. Wu H, Chang D, Huang C (2006) Targeted therapy for cancer. J Cancer Mol 2(2):57–66
5. Faivre S, Djelloul S, Raymond E (2006) New paradigms in anticancer therapy: targeting multiple signaling pathways with kinase inhibitors. Semin Oncol 33:407–420
6. Gil J, Stembalska A, Pesz KA, Sasiadek MM (2008) Cancer stem cells: the theory and perspectives in cancer therapy. J Appl Genet 49(2):193–199
7. Schumacher T, Bunse L, Pusch S, Sahm F, Wiestler B, Quandt J, Menn O, Osswald M, Oezen I, Ott M, Keil M, Balb J, Rauschenbach K, Grabowska AK, Vogler I, Diekmann J, Trautwein N, Eichmuller SB, Okun J, Stevanovic S, Riemer AB, Sahin U, Friese MA, Beckhove P, von Deimling A, Wick W, Platten M (2014) A vaccine-targeting mutant IDH1 induces antitumour immunity. Nature 512(7514):324–327
8. Kleinsmith LJ (2006) Principles of cancer biology. Pearson Benjamin Cummings, San Francisco
9. Singh SK, Hawkins C, Clarke ID, Squire JA, Bayani J, Hide T, Henkelman RM, Cusimano MD, Dirks PB (2004) Identification of human brain tumour initiating cells. Nature 432(7015):396–401
10. Garcia MA, Carrasco E, Ranirez A, Jimenez G, Lopez-Ruiz E, Peran M et al (2012) Apoptosis as a therapeutic target in cancer stem cells: novel strategies and future perspectives. In: Ntuli TM (ed) Apoptosis and medicine. Intech, New York (Chapter 5)
11. Dabelsteen E (1996) Cell surface carbohydrates as prognostic markers in human carcinomas. J Pathol 179:513–516
12. Kure S, Matsuda Y, Hagio M, Ueda J, Naito Z, Ishiwate T (2012) Expression of cancer stem cell markers in pancreatic intraepithelial neoplasias and pancreatic ductal adenocarcinoma. Int J Oncol 41(4):1314–1324

13. Jamieson KH (2008) Chronic myeloid leukemia stem cells. Am Soc Hematol:436–442
14. Wu JC (2014) Immunologic targeting of the cancer stem cells. Stem book, pp 1–37. www.stembook.org/node/529
15. Jones RJ, Armstrong SA (2008) Cancer stem cells in hematopoietic malignancies. Biol Blood Marrow Transplant 14:12–16
16. Walter RB, Appelbaum RA, Estey HE, Bernstein ID (2012) Acute myeloid leukemia stem cells and CD33-targeted immunotherapy. Blood 119:6198–6208
17. Hudziak RM, Lewis GD, Winget M, Fendly BM, Shepard HM, Ullrich A (1989) p185HER2 monoclonal antibody has antiproliferative effects in vitro and sensitizes human breast tumor cells to tumor necrosis factor. Mol Cell Biol 9:1165–1172
18. Horton SJ, Huntly BJ (2012) Recent advances in acute myeloid leukemia stem cell biology. Haematologica 97(7):966–974
19. Carter P, Presta L, Gorman CM, Ridgway JB, Henner D, Wong WL, Rowland AM, Kotts C, Carver ME, Shepard HM (1992) Humanization of an anti-p185HER2 antibody for human cancer therapy. Proc Natl Acad Sci U S A 89(10):4285–4289
20. Martins AM, Vunjak-Novakovic G, Rl R (2014) The current status of iPs cells in cardiac research and their potential for tissue engineering and regenerative medicine. Stem Cell Rev Rep 10:177–190
21. Press OW, Appelbaum F, Ledbetter JA, Martin PJ, Zarling J, Kidd P, Thomas ED (1987) Monoclonal antibody 1F5 (anti-CD20) serotherapy of human B cell lymphomas. Blood 69:584–591
22. Reff ME, Carner K, Chambers KS, Chinn PC, Leonard JE, Raab R, Newman RA, Hanna N, Anderson DR (1994) Depletion of B cells in vivo by a chimeric mouse human monoclonal antibody to CD20. Blood 83:435–445
23. Maloney DG, Liles TM, Czerwinski DK, Waldichuk C, Rosenberg J, Grillo-Lopez A, Levy R (1994) Phase I clinical trial using escalating single-dose infusion of chimeric anti-CD20 monoclonal antibody (IDEC-C2B8) in patients with recurrent B-cell lymphoma. Blood 84:2457–2466
24. Czuczman MS, Grillo-Lopez AJ, White CA, Saleh M, Gordon L, LoBuglio AF, Jonas C, Klippenstein D, Dallaire B, Varns C (1999) Treatment of patients with low-grade B-cell lymphoma with the combination of chimeric anti-CD20 monoclonal antibody and CHOP chemotherapy. J Clin Oncol 17:268–276
25. Slamon DJ, Clark GM, Wong SG, Levin WJ, Ullrich A, McGuire WL (1987) Human breast cancer: correlation of relapse and survival with amplification of the HER-2/neu oncogene. Science 235:177–182
26. Naito K, Takeshita A, Shigeno K, Nakamura S, Fujisawa S, Shinjo K, Yoshida H, Ohnishi K, Mori M, Terakawa S, Ohno R (2000) Calicheamicin-conjugated humanized anti-CD33 monoclonal antibody (gemtuzumab zogamicin, CMA-676) shows cytocidal effect on CD33-positive leukemia cell lines, but is inactive on P-glycoprotein-expressing sublines. Leukemia 14:1436–1443
27. Boghaert ER, Khandke K, Sridharan L, Armellino D, Dougher M, Dijoseph JF, Kunz A, Hamann PR, Sridharan A, Jones S, Discafani C, Damle NK (2006) Tumoricidal effect of calicheamicin immuno-conjugates using a passive targeting strategy. Int J Oncol 28:675–684
28. Genentech Inc. Herceptin® (trastuzumab) (1998, Sept) Investigator brochure. San Francisco
29. Lozanski G, Heerema NA, Flinn IW, Smith L, Harbison J, Webb J, Moran M, Lucas M, Lin T, Hackbarth ML, Proffitt JH, Lucas D, Grever MR, Byrd JC (2004) Alemtuzumab is an effective therapy for chronic lymphocytic leukemia with p53 mutations and deletions. Blood 103:3278–3281
30. Kaufman DB, Leventhal JR, Gallon LG, Parker MA (2006) Alemtuzumab induction and prednisone-free maintenance immunotherapy in simultaneous pancreas-kidney transplantation comparison with rabbit anti-thymocyte globulin induction long term results. Am J Transplant 6:331–339
31. Lundin J, Osterborg A, Brittinger G, Crowther D, Dombret H, Engert A, Epenetos A, Gisselbrecht C, Huhn D, Jaeger U, Thomas J, Marcus R, Nissen N, Poynton C, Rankin E,

Stahel R, Uppenkamp M, Willemze R, Mellstedt H (1998) CAMPATH-1H monoclonal antibody in therapy for previously treated low-grade non-Hodgkin's lymphomas: a phase II multicenter study. European Study Group of CAMPATH-1H Treatment in Low-Grade Non-Hodgkin's Lymphoma. J Clin Oncol 16:3257–3263

32. Chinn PC, Leonard JE, Rosenberg J, Hanna N, Anderson DR (1999) Preclinical evaluation of 90Y-labeled anti-CD20 monoclonal antibody for treatment of non-Hodgkin's lymphoma. Int J Oncol 15:1017–1025

33. Cheson BD (2003) Radioimmunotherapy of non-Hodgkin lymphomas. Blood 101:391–398

34. Nademanee A, Forman S, Molina A, Fung H, Smith D, Dagis A, Kwok C, Yamauchi D, Anderson AL, Falk P, Krishnan A, Kirschbaum M, Kogut N, Nakamura R, O'Donnell M, Parker P, Popplewell L, Pullarkat V, Rodriguez R, Sahebi F, Smith E, Snyder D, Stein A, Spielberger R, Zain J, White C, Raubitschek A (2005) A phase 1/2 trial of high-dose yttrium-90-ibritumomab tiuxetan in combination with high-dose etoposide and cyclophosphamide followed by autologous stem cell transplantation in patients with poor-risk or relapsed non-Hodgkin lymphoma. Blood 106:2896–2902

35. Kaminski MS, Zasadny KR, Francis IR, Fenner MC, Ross CW, Milik AW, Estes J, Tuck M, Regan D, Fisher S, Glenn SD, Wahl RL (1996) Iodine-131-anti-B1 radioimmunotherapy for B-cell lymphoma. J Clin Oncol 14:1974–1981

36. Vose JM, Wahl RL, Saleh M, Rohatiner AZ, Knox SJ, Radford JA, Zelenetz AD, Tidmarsh GF, Stagg RJ, Kaminski MS (2000) Multicenter phase II study of iodine-131 tositumomab for chemotherapy relapsed/refractory low-grade and transformed low-grade B-cell non-Hodgkin's lymphomas. J Clin Oncol 18:1316–1323

37. Liu SY, Eary JF, Petersdorf SH, Martin PJ, Maloney DG, Appelbaum FR, Matthews DC, Bush SA, Durack LD, Fisher DR, Gooley TA, Bernstein ID, Press OW (1998) Follow-up of relapsed B-cell lymphoma patients treated with iodine-131-labeled anti-CD20 antibody and autologous stem cell rescue. J Clin Oncol 16:3270–3278

38. Huang SM, Bock JM, Harari PM (1999) Epidermal growth factor receptor blockade with C225 modulates proliferation, apoptosis, and radiosensitivity in squamous cell carcinomas of the head and neck. Cancer Res 59:1935–1940

39. Baselga J, Pfister D, Cooper MR, Cohen R, Burtness B, Bos M, D'Andrea G, Seidman A, Norton L, Gunnett K, Falcey J, Anderson V, Waksal H, Mendelsohn J (2000) Phase I studies of anti-epidermal growth factor receptor chimeric antibody C225 alone and in combination with cisplatin. J Clin Oncol 18:904–914

40. Jimeno A, Rubio-Viqueira B, Amador ML, Oppenheimer D, Bouraoud N, Kulesza P, Sebastiani V, Maitra A, Hidalgo M (2005) Epidermal growth factor receptor dynamics influences response to epidermal growth factor receptor targeted agents. Cancer Res 65:3003–3010

41. Willett CG, Boucher Y, di Tomaso E, Duda DG, Munn LL, Tong RT, Chung DC, Sahani DV, Kalva SP, Kozin SV, Mino M, Cohen KS, Scadden DT, Hartford AC, Fischman AJ, Clark JW, Ryan DP, Zhu AX, Blaszkowsky LS, Chen HX, Shellito PC, Lauwers GY, Jain RK (2004) Direct evidence that the VEGF-specific antibody bevacizumab has antivascular effects in human rectal cancer. Nat Med 10:145–147

42. Ferrara N, Hillan KJ, Gerber HP, Novotny W (2004) Discovery and development of bevacizumab, an anti-VEGF antibody for treating cancer. Nat Rev Drug Discov 3:391–400

43. Sonpavde G (2004) Bevacizumab in colorectal cancer. N Engl J Med 351:1690–1691

44. Kabbinavar FF, Schulz J, McCleod M, Patel T, Hamm JT, Hecht JR, Mass R, Perrou B, Nelson B, Novotny WF (2005) Addition of bevacizumab to bolus fluorouracil and leucovorin in first-line metastatic colorectal cancer: results of a randomized phase II trial. J Clin Oncol 23:3697–3705

45. Mayfield J, Pavlovic M (2014) A concept of cancer stem cells: the current understanding and a look ahead. ART 41(1–2):49–60

46. Bonnet D, Dick JE (1997) Human acute myeloid leukemia is organized as a hierarchy that originates from a primitive hematopoietic cell. Nat Med 3:730–737

47. Al-Hajj M, Wicha MS, Benito-Hernandez A, Morrison SJ, Clarke MF (2003) Prospective identification of tumorigenic breast cancer cells. Proc Natl Acad Sci U S A 100:3983–3988

48. Collins AT, Berry PA, Hyde C, Stower MJ, Maitland NJ (2005) Prospective identification of tumorigenic prostate cancer stem cells. Cancer Res 65:10946–10951
49. Li C, Heidt DG, Dalerba P, Burant CF, Zhang L, Adsay V, Wicha M, Clarke MF, Simeone DM (2007) Identification of pancreatic cancer stem cells. Cancer Res 67:1030–1037
50. O'Brien CA, Pollett A, Gallinger S, Dick JE (2007) A human colon cancer cell capable of initiating tumour growth in immunodeficient mice. Nature 445:106–110
51. Patrawala L, Calhoun T, Schneider-Broussard R, Li H, Bhatia B, Tang S, Reilly JG, Chandra D, Zhou J, Claypool K, Coghlan L, Tang DG (2006) Highly purified CD44 prostate cancer cells from xenograft human tumors are enriched in tumorigenic and metastatic progenitor cells. Oncogene 25:1696–1708
52. Prince ME, Sivanandan R, Kaczorowski A, Wolf GT, Kaplan MJ, Dalerba P, Weissman IL, Clarke MF, Ailles LE (2007) Identification of a subpopulation of cells with cancer stem cell properties in head and neck squamous cell carcinoma. Proc Natl Acad Sci U S A 104:973–978
53. Ricci-Vitiani L, Lombardi DG, Pilozzi E, Biffoni M, Todaro M, Peschle C, De Maria R (2007) Identification and expansion of human colon-cancer-initiating cells. Nature 445:111–115
54. Gaitanis A, Staal S (2010) Liposomal doxorubicin and nab-paclitaxel: nanoparticle cancer chemotherapy in current clinical use. Methods Mol Biol 624:385–392
55. Graishar WJ, Tjulndin S, Davidson N, Shaw H, Desai N, Bhar P et al (2005) Phase III trial of nanoparticle albumin-bound paclitaxel compared with polyethylated castor oil-based paclitaxel in women with breast cancer. J Clin Oncol 23(31):7794–7803
56. Farokhzad OC, Jon S, Khademhosseini A, Tran TNT, LaVan DA, Langer R (2004) Nanoparticle-aptamer bioconjugates a new approach for targeting prostate cancer cells. Cancer Res 64(21):7668–7672
57. Kirson ED, Gurvich Z, Schneiderman R, Dekel E, Itzhaki A, Wasserman Y et al (2004) Disruption of cancer cell replication by alternating electric fields. Cancer Res 64(9):3288–3295
58. Stupp R, Wong ET, Kanner AA, Steinberg D, Engelhard H, Heidecke V et al (2012) NovoTTF-100A versus physician's choice chemotherapy in recurrent glioblastoma: a randomised phase III trial of a novel treatment modality. Eur J Cancer 48(14):2192–2202
59. Dylla SJ, In-Kyung P, Gurney AL (2009) Emerging technology platforms for stem cells. Wiley, Hoboken
60. Pavlovic M, Balint B (2013) Stem cells and tissue engineering. Springer, New York
61. Reya T, Morrison SJ, Clarke MF, Weissman IL (2001) Stem cells, cancer, and cancer stem cells. Nature 414(6859):105–111
62. Fang D, Nguyen TK, Leishear K, Finko R, Kulp AN, Hotz S, Van Belle PA, Xu X, Elder DE, Herlyn M (2005) A tumorigenic subpopulation with stem cell properties in melanomas. Cancer Res 65:9328–9337
63. Dalerba P, Dylla SJ, Park IK, Liu R, Wang X, Cho RW, Hoey T, Gurney A, Huang EH, Simeone DM, Shelton AA, Parmiani G, Castelli C, Clarke MF (2007) Phenotypic characterization of human colorectal cancer stem cells. Proc Natl Acad Sci U S A 104:10158–10163
64. Kim CF, Jackson EL, Woolfenden AE, Lawrence S, Babar I, Vogel S, Crowley D, Bronson RT, Jacks T (2005) Identification of bronchioalveolar stem cells in normal lung and lung cancer. Cell 121:823–835
65. Gibbs CP, Kukekov VG, Reith JD, Tchigrinova O, Suslov ON, Scott EW, Ghivizzani SC, Ignatova TN, Steindler DA (2005) Stem-like cells in bone sarcomas: implications for tumorigenesis. Neoplasia 7:967–976

Chapter 10
Targeted Cancer Stem Cell Therapy

There is always a REASON, a 'why', for someone getting cancer. How in the world can anyone truly get better without discovering exactly what that IS.

Dr. Conners

General: Remote Control of Targeted Drug Delivery as Possibility for Cancer Stem Cell Therapy

About 15 years ago, MIT professors Robert Langer and Michael Cima came up with interesting idea to develop a programmable, wirelessly controlled microchip that would deliver drugs after implantation in a patient's body. Finally, the MIT researchers and scientists from MicroCHIPS Inc. reported that they have successfully used such a chip to administer daily doses of an osteoporosis drug normally given by injection.

According to Dr. Langer, the results represent the first successful test of such a device and could help user in a new era of telemedicine—delivering health care over a distance. That means literally have a pharmacy on a chip. According to these scientists, you can do remote control delivery, you can do pulsatile drug delivery, and you can deliver multiple drugs.

In the new study, funded and overseen by MicroCHIPS, scientists used the programmable implants to deliver an osteoporosis drug called teriparatide to seven women aged 65–70. The study found that the device delivered dosages comparable to injections, and there were no adverse side effects. These programmable chips could dramatically change treatment not only for osteoporosis, but also for many other diseases, including cancer and multiple sclerosis. The scientists say that patients with chronic diseases, regular pain-management needs, or other conditions that require frequent or daily injections could benefit from this technology. This also avoids the compliance issue completely, points to a future where you have fully automated drug regimens, and approach is much more comfortable for the patient.

© The Author(s) 2015
M. Pavlovic, B. Balint, *Bioengineering and Cancer Stem Cell Concept*,
DOI 10.1007/978-3-319-25670-2_10

In 1999, the MIT team published its initial findings in Nature, and MicroCHIPS was founded and licensed the microchip technology from MIT. It was a *solid-state silicon microchip* that can provide controlled release of single or multiple chemical substances on demand [1]. The release mechanism was based on the electrochemical dissolution of thin anode membranes covering microreservoirs filled with chemicals in solid, liquid, or gel form [1]. The company refined the chips, including adding a hermetic seal and a release system that works reliably in living tissue. Teriparatide is a polypeptide and therefore much less chemically stable than small-molecule drugs, so sealing it hermetically to preserve it was an important achievement. The human clinical trial began in Denmark in January 2011. Chips were implanted during a 30-min procedure at a doctor's office using local anesthetic, and remained in the patients for 4 months. The implants proved safe, and patients reported they often forgot they even had the implant.

Another approach is the work on nanoparticles with a reservoir under remote control in order to control the drug delivery into targeted tissues or organs.

If analyzed, it seems that recently, the most interesting examples are coming from MIT (Langer's group) and the group of Dr. Sakhrat Khizroev from FIU [1–3].

1. A reservoir that could be remotely triggered to release a drug would enable the patient or physician to achieve on-demand, reproducible, repeated, and tunable dosing [1]. Devices that release a drug in response to a remote trigger would enable on-demand control of the timing and dose of drug released. They would allow the patient or physician to adjust therapy precisely to a target effect, thus improving treatment and reducing toxicity. Langer's group has developed implantable reservoirs that release a drug when irradiated with near-infrared laser light. The release rate was correlated to laser intensity, with negligible leakage between doses. Devices containing aspart, a fast-acting analog of insulin, were implanted in diabetic rats and were able to achieve glycemic control upon irradiation. Such devices can be loaded with a wide range of drugs to treat a variety of clinical indications [1].

2. Dr. Khizroev's group has fabricated magnetic nanoparticles that have targeted ovarian carcinoma cells, controlled by a low energy remote magnetic field. They confirmed through kinetic studies that the drug penetrated through the tumor cell membrane and eradicated the majority of tumor cells within a 24 h period without affecting surrounding healthy cells [3]. It would be very interesting to target cancer stem cells only in order to see how does the concept of targeted Cancer Stem Cell Therapy work. The future work will show how relevant these two interesting approaches are for therapeutic purposes.

Nanotechnology/Nanoparticles

Nanoparticles are particles between 1 and 100 nm in size. In nanotechnology, a particle is defined as a small object that behaves as a whole unit with respect to its transport and properties. Particles are further classified according to diameter [4].

Currently many substances are under investigation for drug delivery and more specifically for cancer therapy [5]. Interestingly, pharmaceutical sciences are using nanoparticles to reduce toxicity and side effects of drugs and up to recently did not realize that carrier systems themselves may impose risks to the patient. For nanoparticles, the knowledge on particle toxicity as obtained in inhalation toxicity shows the way how to investigate the potential hazards of nanoparticles. The toxicology of particulate matter differs from toxicology of substances as the composing chemical(s) may or may not be soluble in biological matrices, thus influencing greatly the potential exposure of various internal organs [5]. This may vary from a rather high local exposure in the lungs and a low or neglectable exposure for other organ systems after inhalation [1–5]. However, absorbed species may also influence the potential toxicity of the inhaled particles. For nanoparticles, the situation is different as their size opens the potential for crossing the various biological barriers within the body. From a positive viewpoint, especially the potential to cross the blood brain barrier may open new ways for drug delivery into the brain. In addition, the nanosize also allows for access into the cell and various cellular compartments including the nucleus. A multitude of substances are currently under investigation for the preparation of nanoparticles for drug delivery, varying from biological substances like albumin, gelatine and phospholipids for liposomes, and more substances of a chemical nature like various polymers and solid metal containing nanoparticles [5–14]. It is obvious that the potential interaction with tissues and cells, and the potential toxicity, greatly depends on the actual composition of the nanoparticle formulation. This paragraph in the chapter provides an overview on some of the currently used systems for nanoparticle drug delivery in cancer and cancer stem cell research [5–14].

Interesting ideas are coming from the work of Dr. Andrew Koehl who indicates an amazing fact that there are compounds on your breath that indicate illness, that's been shown through a number of studies and we can detect those with for that purpose designed nanosensors [6]. There have already been a number of research papers published suggesting we can detect cancer, tuberculosis, and asthma [6].

Anticancer drugs can often shrink tumors but don't kill cancer stem cells (CSCs). Although CSCs might only make up a small part of a tumor, their resistance to drugs allows them to persist. They can then cause a tumor to regrow or spread cancerous cells throughout the body. Xiaoming He and colleagues wanted to develop a nanoparticle system to overcome these cells' defenses by placing anticancer drug doxorubicin into nanoparticles coated with chitosan, a natural polysaccharide that can specifically target CSCs [7]. Once in the acidic environment of the tumor, the nanoparticles degraded and released the drug. Tests on tiny, tissue-like clumps of both normal and cancer stem cells in vitro and on human breast tumors grown in mice showed the therapy successfully killed CSCs and destroyed tumors. The mice showed no obvious side effects. The group has shown that chitosan binds with a receptor CD44 on cancer stem-like cells, enabling the nanoparticles to target the malignant stem-like cells in a tumor [7]. The magnetic beads for extraction of the CSCs are already designed. This will enable the extraction of each particular CSCs and tests for targeted therapy in vitro. All other nanoparticles will probably be more precisely tested and used as targeted or eliminated [15–44].

Biomagnetism/Magnetic Particles

This approach is described in precious Chap. 9.

Immunotherapy (Rational Vaccine Design) for CSCs

Despite the similar name, CSCs are distinct from ESCs, and the two avenues of research are separate. Thus, CSCs are tumor cells that remain present, and ultimately resistant, after chemotherapy or radiation treatment. Scientists disagree on whether these cells have unique properties, but those who support the uniqueness idea have argued that these cells regenerate the tumors that lead to relapse [45].

There are several immunological approaches in cancer treatment, but novel approach is the rational vaccine design (RVD) for cancer against CSCs [46, 47]. Due to possibility to extract and purify CSCs only, it is also possible to develop the vaccine against these cells [46]. Maybe it would have been the best possible approach if the CSC theory really works. Thus, Li and colleagues extracted and purified cancer stem cells from two immunocompetent mouse models and used them to prepare the vaccine [47]. They found that these enriched cancer stem cells were immunogenic and far more effective as an antigen source compared with the unselected tumor cells normally used in previous immunotherapy trials [47]. The mechanistic investigations found that when antibodies were primed with CSCs, they were capable of targeting CSCs and conferring antitumor immunity. The researchers also found that cytotoxic T lymphocytes harvested from cancer stem cell-vaccinated hosts were capable of killing cancer stem cells in vitro [47]. Is it possible that dendritic cell-based vaccine (DC) would work even better or in combination with antibodies?

Epigenetic Synthetic Lethal Therapeutic Approaches to ARID1A Mutated Cancers

Rugang Zhang, Ph.D., from Wistar Institute is the name linked to a major discovery of recent cancer genome-wide sequencing studies, e.g., the identification of significant changes in genes responsible for modifying chromatin structure [48]. ARID1A, an epigenetic chromatin remodeler, is known to be among the genes that show the highest mutation rates across many cancer types. Notably, ARID1A is mutated in over 50 % of ovarian clear cell carcinomas, which currently has no effective therapy. However, targeted cancer therapy based on ARID1A mutational status has not been described. Here we show that several epigenetic inhibitors act in a synthetic lethal manner in ARID1A mutated ovarian cancer cells. ARID1A mutational status correlates with response to these inhibitors. Significantly, these inhibitors cause the regression of ARID1A mutated but not wild type ovarian tumors in vivo [48]. Together, these data indicate that epigenetic synthetic lethality using pharmacological inhibitors represents a novel treatment strategy for ARID1A mutated cancers.

Genetic Engineering-Based Therapy: Engineered Normal (Toxic) Stem Cells as a Therapeutic Tool for Cancer

Quite recently, thanks to developed bioengineering methodology, scientists engineered *toxin-secreting stem cells* to treat brain tumors (October 24, 2014). Proof-of-concept study highlights new therapeutic use of engineered human stem cells. Harvard Stem Cell Institute scientists at Massachusetts General Hospital have devised a new way to use stem cells in the fight against brain cancer. A team led by neuroscientist Khalid Shah, who recently demonstrated the value of stem cells loaded with cancer-killing herpesviruses, now has a way to genetically engineer stem cells so that they can produce and secrete tumor-killing toxins [49, 50].

This method solves the delivery issue that probably led to the failure of recent clinical trials aimed at delivering purified cancer-killing toxins into patient's brain. These stem cells are placed at the site encapsulated in a biodegradable gel. Cancer-killing toxins have been used with great success in a variety of blood cancers, but they don't work as well in solid tumors because the cancers aren't as accessible and the toxins have a short half-life [50]. A few years ago, this group has recognized that stem cells could be used to continuously deliver these therapeutic toxins to tumors in the brain, but first the group needed to genetically engineer stem cells that they could resist being killed themselves by the toxins. Now, they had toxin-resistant stem cells that can make and release cancer-killing drugs. Cytotoxins are deadly to all cells, but since the late 1990s, researchers have been able to tag toxins in such a way that they *only enter cancer cells* with specific surface molecules, making it possible to get a toxin into a cancer cell without posing a risk to normal cells. Once inside of a cell, the toxin disrupts the cell's ability to make proteins and, within days, the cell starts to die. Shahâ's stem cells escape this fate because they are made with a mutation that doesn't allow the toxin to act inside the cell. The *toxin-resistant stem cells* also have an *extra bit of genetic code that allows them to make and secrete the toxins* [50]. Any cancer cells that these toxins encounter do not have this natural defense and therefore die. Shah and his team *induced toxin resistance* in human neural stem cells and subsequently engineered them to produce targeted toxins. The team tested these stem cells in a clinically relevant mouse model of brain cancer, where you resect the tumors and then implant the stem cells encapsulated in a gel into the resection cavity. After doing all of the molecular analysis and imaging to track the inhibition of protein synthesis within brain tumors, they did see the toxins kill the cancer cells and eventually prolonging the survival in animal models of resected brain tumors. Next plans involve intention to rationally combine the toxin-secreting stem cells with a number of different therapeutic stem cells developed by this team to further enhance their positive results in mouse models of glioblastoma, the most common brain tumor in human adults [50]. The group predicts that they will bring these therapies into clinical trials within the next 5 years within the projects such as:

- Leukemia stem cells and targeted therapy
- Breast CSCs and targeted therapy
- Prostate CSCs and targeted therapy
- Glioma stem cells and targeted therapy
- Colorectal stem cells and targeted therapy
- Targeted therapy against other CSCs

Conclusions

It was inevitable for CSC concept and therapy to advance. Explosive development of nanotechnology, magnetoparticle application, technologies for CSCs extraction (magnetic beads) and RVD and other approaches have placed CSC concept into the focus of cancer research. It is also inevitable that the concept is being re-investigated from many point of views and being confirmed with every single new day. There is a great hope that this new target in cancer treatment will successfully replace radiation and chemotherapy approaches with unwanted and sometimes deleterious side effects [51–56].

References

1. Santini JT Jr, Cima MJ, Langer R (1999) A controlled-release microchip. Nature 397:335–338
2. Yang T, Rycaj K (2015) Targeted therapy against cancer stem cells (review). Oncol Lett 10:27–33
3. Timko BP, Arruebo M, Shankarappa SA, McAlvin JB, Okonkwo OS, Mizrahi B, Stefanescu CF, Gomez L, Zhu J, Zhu A, Santamaria J, Langer R, Kohane DS (2014) Near infrared-actuated devices for remotely controlled drug delivery. Proc Natl Acad Sci U S A 111(4):1349–1354
4. Guduru R, Liang P, Runowicz C, Nair M, Atluri V, Khizroev S (2013) Magneto-electric nanoparticles (MENs) to enable field-controlled high-specificity drug delivery to eradicate ovarian cancer cells. Sci Rep 3:2953
5. De Jong WH, De Borm PJA (2008) Drug delivery and nanoparticles: applications and hazards. Int J Nanomedicine 3(2):133–149
6. Andrew Koehl http://www.owlstonenanotech.com/company/management
7. Rao W, Wang H, Han J, Zhao S, Dumbleton J, Agarwal P, Zhang W, Zhao G, Yu J, Zynger DL, Lu X, He X (2015) Chitosan-decorated doxorubicin-encapsulated nanoparticle targets and eliminates tumor reinitiating cancer stem-like cells. ACS Nano 9(6):5725. doi:10.1021/nn506928p
8. Cirillo G, Iemma F, Puoci F, Parisi OI, Curcio M, Spizzirri UG, Picci N (2009) Imprinted hydrophilic nanospheres as drug delivery systems for 5-fluorouracil sustained release. J Drug Target 17(1):72–77

9. Cirillo G, Hampel S, Spizzirri GU, Paris O, Picci N, Iemma F (2014) Carbon nanotubes hybrid hydrogels in drug delivery: a perspective review. Biomed Res Int 2014:825017

10. Daum N, Tscheka C, Neumeyer A, Schneider M (2012) Novel approaches for drug delivery systems in nanomedicine: effects of particle design and shape. Wiley Interdiscip Rev Nanomed Nanobiotechnol 4(1):52–65

11. Brannon-Peppas L, Blanchette JO (2004) Nanoparticle and targeted systems for cancer therapy. Adv Drug Deliv Rev 56(11):1649–1659

12. Davis ME (2008) Nanoparticle therapeutics: an emerging treatment modality for cancer. Nat Rev Drug Discov 7(9):771–782

13. Farokhzad OC et al (2004) Nanoparticle-aptamer bioconjugates a new approach for targeting prostate cancer cells. Cancer Res 64(21):7668–7672

14. Hapira A, Livney YD, Broxterman HJ, Assaraf YG (2011) Nanomedicine for targeted cancer therapy: towards the overcoming of drug resistance. Drug Resist Updat 14:150–163

15. Gil J, Stembalska A, Pesz KA, Sasiadek MM (2008) Cancer stem cells: the theory and perspectives in cancer therapy. J Appl Genet 49(2):193–199

16. Gradishar WJ, Tjulandin S, Davidson N, Shaw H, Desai N, Bhar P, Hawkins M, O'Shaughnessy J (2005) Phase III trial of nanoparticle albumin-bound paclitaxel compared with polyethylated castor oil-based paclitaxel in women with breast cancer. J Clin Oncol 23(31):7794–7803

17. Hapira A, Livney YD, Broxterman HJ, Assaraf YG (2011) Nanomedicine for targeted cancer therapy: towards the overcoming of drug resistance. Drug Resist Updat 14:150–163

18. Ahmed N, Fessi H, Elaissari A (2012) Theranostic applications of nanoparticles in cancer. Drug Discov Today 17(17–18):928–934

19. Vinogradov S, Wei X (2012) Cancer stem cells and drug resistance: the potential of nanomedicine. Nanomedicine (Lond) 7:597–615

20. Lim KJ, Bisht S, Bar EE, Maitra A, Eberhart CG (2011) A polymeric nanoparticle formulation of curcumin inhibits growth, clonogenicity and stem-like fraction in malignant brain tumors. Cancer Biol Ther 11:464–473

21. Zhou BBS, Zhang H, Damelin M, Geles KG, Grindley JC et al (2009) Tumour-initiating cells: challenges and opportunities for anticancer drug discovery. Nat Rev Drug Discov 8:806–823

22. McDermott SP, Wicha MS (2010) Targeting breast cancer stem cells. Mol Oncol 4:404–419

23. Wu X, Chen H, Wang X (2012) Can lung cancer stem cells be targeted for therapies? Cancer Treat Rev 38:580–588

24. Mamaeva V, Rosenholm JM, Bate-Eya LT, Bergman L, Peuhu E et al (2011) Mesoporous silica nanoparticles as drug delivery systems for targeted inhibition of Notch signaling in cancer. Mol Ther 19:1538–1546

25. Yang ZF, Ho DW, Ng MN, Lau CK, Yu WC et al (2008) Significance of CD90+ cancer stem cells in human liver cancer. Cancer Cell 13:153–166

26. Liu C, Zhao G, Liu J, Ma N, Chivukula P et al (2009) Novel biodegradable lipid nano complex for siRNA delivery significantly improving the chemosensitivity of human colon cancer stem cells to paclitaxel. J Control Release 140:277–283

27. Bader AG, Brown D, Stoudemire J, Lammers P (2011) Developing therapeutic microRNAs for cancer. Gene Ther 18:1121–1126

28. Piao L, Zhang M, Datta J, Xie X, Su T et al (2012) Lipid-based nanoparticle delivery of pre-miR-107 inhibits the tumorigenicity of head and neck squamous cell carcinoma. Mol Ther 20:1261–1269

29. Yin D, Ogawa S, Kawamata N, Leiter A, Ham M et al (2013) miR-34a Functions as a tumor suppressor modulating EGFR in glioblastoma multiforme. Oncogene 32(9):1155–1163

30. Ugras S, Brill E, Jacobsen A, Hafner M, Socci ND et al (2011) Small RNA sequencing and functional characterization reveals MicroRNA-143 tumor suppressor activity in liposarcoma. Cancer Res 71:5659–5669

31. Pramanik D, Campbell NR, Karikari C, Chivukula R, Kent OA et al (2011) Restitution of tumor suppressor microRNAs using a systemic nanovector inhibits pancreatic cancer growth in mice. Mol Cancer Ther 10:1470–1480

32. Issels RD (2008) Hyperthermia adds to chemotherapy. Eur J Cancer 44:2546–2554
33. Hirsch LR, Stafford RJ, Bankson JA, Sershen SR, Rivera B et al (2003) Nanoshell-mediated near-infrared thermal therapy of tumors under magnetic resonance guidance. Proc Natl Acad Sci U S A 100:13549–13554
34. Yang K, Zhang S, Zhang G, Sun X, Lee ST et al (2010) Graphene in mice: ultrahigh in vivo tumor uptake and efficient photothermal therapy. Nano Lett 10:3318–3323
35. Kam NW, O'Connell M, Wisdom JA, Dai H (2005) Carbon nanotubes as multifunctional biological transporters and near-infrared agents for selective cancer cell destruction. Proc Natl Acad Sci U S A 102:11600–11605
36. Burke A, Ding X, Singh R, Kraft RA, Levi-Polyachenko N et al (2009) Long-term survival following a single treatment of kidney tumors with multiwalled carbon nanotubes and near-infrared radiation. Proc Natl Acad Sci U S A 106:12897–12902
37. Ding X, Singh R, Burke A, Hatcher H, Olson J et al (2011) Development of iron-containing multiwalled carbon nanotubes for MR-guided laser-induced thermotherapy. Nanomedicine (Lond) 6:1341–1352
38. Huang X, El-Sayed IH, Qian W, El-Sayed MA (2006) Cancer cell imaging and photothermal therapy in the near-infrared region by using gold nanorods. J Am Chem Soc 128:2115–2120
39. Burke AR, Singh RN, Carroll DL, Wood JC, D'Agostino RB et al (2012) The resistance of breast cancer stem cells to conventional hyperthermia and their sensitivity to nanoparticle-mediated photothermal therapy. Biomaterials 33:2961–2970
40. Atkinson RL, Zhang M, Diagaradjane P, Peddibhotla S, Contreras A et al (2010) Thermal enhancement with optically activated gold nanoshells sensitizes breast cancer stem cells to radiation therapy. Sci Transl Med 2:55–79
41. Phillips TM, McBride WH, Pajonk F (2006) The response of CD24(–/low)/CD44+ breast cancer-initiating cells to radiation. J Natl Cancer Inst 98:1777–1785
42. Galanzha EI, Kim JW, Zharov VP (2009) Nanotechnology-based molecular photoacoustic and photothermal flow cytometry platform for in vivo detection and killing of circulating cancer stem cells. J Biophotonics 2:725–735
43. Galanzha EI, Shashkov EV, Kelly T, Kim JW, Yang L et al (2009) In vivo magnetic enrichment and multiplex photoacoustic detection of circulating tumour cells. Nat Nanotechnol 4:855–860
44. Kim JW, Galanzha EI, Shashkov EV, Moon HM, Zharov VP (2009) Golden carbon nanotubes as multimodal photoacoustic and photothermal high-contrast molecular agents. Nat Nanotechnol 4:688–694
45. Hillner BE, Smith TJ (1991) Efficacy and cost effectiveness of adjuvant chemotherapy in women with node-negative breast cancer: a decision-analysis model. N Engl J Med 324(3):160–168
46. Ning N, Pan Q, Zheng F, Teitz-Tennenbaum S, Egenti M, Yet J, Li M, Ginestier C, Wicha MS, Moyer JS, Prince MEP, Xu Y, Zhang X-L, Huang S, Chang AE, Li Q (2012) Cancer stem cell vaccination confers significant antitumor immunity. Cancer Res 72(7):1853. doi:10.1158/0008-5472.CAN-11-1400
47. Xia X, Mai J, Wang R, Shen H (2015) Porous silicon microparticle potentiates anti-tumor immunity by enhancing cross-presentation and inducing type I interferon response. Cell Rep 11:957–966
48. Biter B, Aird KAM, Garipov A, Li H, Amatangelo M, Zhang R et al (2015) Targeting EZH2 methyltransferase activity in ARID1A mutated cancer cells is synthetic lethal. Nat Med 21(3):231–238
49. Stupp R, Wong ET, Kanner AA, Steinberg D, Engelhard H, Heidecke V, Kirson ED, Taillibert S, Liebermann F, Dbalý V, Ram Z, Villano JL, Rainov N, Weinberg U, Schiff D, Kunschner L, Raizer J, Honnorat J, Sloan A, Malkin M, Landolfi JC, Payer F, Mehdorn M, Weil RJ, Pannullo SC, Westphal M, Smrcka M, Chin L, Kostron H, Hofer S, Bruce J, Cosgrove R, Paleologous N, Palti Y, Gutin PH (2012) NovoTTF-100A versus physician's choice chemotherapy in recurrent glioblastoma: a randomised phase III trial of a novel treatment modality. Eur J Cancer 48(14):2192–2202

50. Stuckey DW et al (2015) Engineering toxin-resistant therapeutic stem cells to treat brain tumors. Stem Cells 33(2):589–600. doi:10.1002/stem.1874
51. al-Sarraf M, Martz K, Herskovic A, Leichman L, Brindle JS, Vaitkevicius VK, Cooper J, Byhardt R, Davis L, Emami B (1997) Progress report of combined chemoradiotherapy versus radiotherapy alone in patients with esophageal cancer: an intergroup study. J Clin Oncol 15(1):277–284
52. Citron ML, Berry DA, Cirrincione C, Hudis C, Winer EP, Gradishar WJ, Davidson NE, Martino S, Livingston R, Ingle JN, Perez EA, Carpenter J, Hurd D, Holland JF, Smith BL, Sartor CI, Leung EH, Abrams J, Schilsky R, Muss HB, Norton L (2003) Randomized trial of dose-dense versus conventionally scheduled and sequential versus concurrent combination chemotherapy as postoperative adjuvant treatment of node-positive primary breast cancer: first report of Intergroup Trial C9741/Cancer and Leukemia Group B Trial 9741. J Clin Oncol 21(8):1431–1439
53. Collins AT, Berry PA, Hyde C, Stower MJ, Maitland NJ (2005) Prospective identification of tumorigenic prostate cancer stem cells. Cancer Res 65(23):10946–10951
54. Delaney G, Jacob S, Featherstone C, Barton M (2005) The role of radiotherapy in cancer treatment. Cancer 104(6):1129–1137
55. Fisher B, Wolmark N, Rockette H, Redmond C, Deutsch M, Wickerham DL, Fisher ER, Caplan R, Jones J, Lerner H et al (1988) Postoperative adjuvant chemotherapy or radiation therapy for rectal cancer: results from NSABP Protocol R-011. J Nat Cancer I 80(1):21–29
56. Gagliardi G, Lax I, Ottolenghi A, Rutqvist LE (1996) Long-term cardiac mortality after radiotherapy of breast cancer—application of the relative seriality model. Br J Radiol 69(825):839–846

Chapter 11
Bioengineered CSC Tumors

A student: Dr. Einstein, Aren't these the same questions as last year's [physics] final exam? Dr. Einstein: Yes; but this year the answers are different.

Albert Einstein

Introduction

It was much easier to design and create the liquid tissue (blood tumors) than a hard tissue-based tumor models. For example, HSCs can be subdivided into long-term self-renewing HSCs, short-term self-renewing HSCs, and multipotent progenitors (red arrows indicate self-renewal). They give rise to common lymphoid progenitors (CLPs; the precursors of all lymphoid cells) and common myeloid progenitors (CMPs; the precursors of all myeloid cells). Both CMPs/GMPs (granulocyte macrophage precursors) and CLPs can give rise to all known mouse dendritic cells (DC).

The isolation of precursors in the hematopoietic system has allowed the generation of a series of mouse models for myeloid leukemia (see box below). The expression of the oncogenes BCL-2, BCR-Abl, and PML-RARa under the control of the hMRP8 promoter, individually or together, and in combination with Fas deficiency, results in diseases that resemble several human leukemias, including chronic myelomonocytic leukemia (CMML), acute myeloid leukemia (AML), acute promyelocytic leukemia (APML), and chronic myeloid leukemia (CML)/Blast.

Leukaemic mouse models	
CMML	MRP8-BCL-2
AML	MRP8-BCL2Xlpr/lpr
APML	MRP8-PML-RARa
CML/Blast	MRP8-BCRablXBCL-2

ErP erythrocyte precursor, *MEP* megakaryocyte erythrocyte precursor, *MkP* megakaryocyte precursor, *NK* natural killer [1]

© The Author(s) 2015
M. Pavlovic, B. Balint, *Bioengineering and Cancer Stem Cell Concept*,
DOI 10.1007/978-3-319-25670-2_11

First Tumor Models Based on CSCs

The initial efforts of John Dick set the stage for an evidence-based concept of cancer stem cells using the NOD/SCID mice model. He and his team successfully transplanted acute myelogenous leukemia (AML) stem cells from human patients into the murine model, wherein the stem cells regenerated AML in the mice [2].

The situation with solid tumor models built up from Cancer Stem Cells was much more difficult given the fact that the solid tumor is far more than a collection of transformed proliferating cells forming tumor mass. They contain cells, extracellular matrix (ECM), T and B lymphocytes, and specific vasculature, the features that liquid tumors do not possess.

As mentioned earlier, at the very beginning, in a study done by Li et al. [3], pancreatic cancer cells were endowed in CD44+CD24+ESA+ cell population representing only 0.2–0.8 % of the total pancreatic cancer cell population. These cells, exhibiting the potential to initiate in vivo pancreatic cancer, are highly concentrated in this phenotypically-defined population since as few as 100 CD44+CD24+ESA+ cells injected in vivo in immunosuppressed mice are forming tumors histologically indistinguishable from human pancreatic cancer tumors from which they originated. The equivalent tumor potential could only be obtained with 100-fold more cells characterized by the CD44-CD24-ESA-phenotype [3, 4] (Chap. 3, Fig. 3.2).

Similar features for human breast cancer cells were shown by Al-Hajj et al. [5]. In this case, the breast cancer cells were endowed by CD44+CD24–/low Lineage-phenotype. This cell population was revealed to be tumorigenic after in vivo injection to immunodeficient mice in the 100 cell dose, which is not the case for the alternate phenotypes even in doses of thousands of cells (26). In similar preparation as Li et al. and their pancreatic cancer model, the 8-week-old NOD/SCID mice were used as the murine model for cancer cell implantation. Again, a xenograft of human breast cancer tissue was taken 1 h after surgical removal. The mincing, washing, and preparation was exactly as Li et al. had done so in the previous model [3, 4] (Chap. 3, Fig. 3.4).

3-D Printing as "Universal" Bioengineering Method

Organ printing offers exciting prospects for fabricating complex three-dimensional (3-D) multicellular organs in vitro by integrating technologies from the fields of engineering, biomaterials science, cell biology, physics, and medicine. It creates well-organized tissue/organ constructs via 3-D printing technology by delivering precisely characterized progenitor cell populations and suitable biomaterials in a defined and organized manner, at a targeted location, in adequate numbers, and within the right environment for repairing or replacing defective human tissues and organs [6]. 3-D printing technology enables the fabrication of complex desired 3-D shapes through a layer-by-layer deposition process and mimicry of the native tissue microenvironment with precisely controlled positioning of ultra-low volumes of biomaterials along with cells on a micro- and/or nanoscale. Various 3D printing systems for organ printing have been developed that can be classified into three major classes: stereo lithography-based, inkjet-based, and dispensing-based 3-D printing. All these 3-D printing methods use computer-aided design/computer-aided manufacturing (CAD/CAM) technology to realize the desired 3-D shape of tissue/ organ constructs. Cells and biomolecules are usually printed by encapsulating them in various dispensable hydrogels, and the mixture is typically called "bioink" [6]. The role of bioink is to protect the cells from harsh environments during the printing process, as well as to enhance cellular functions, such as proliferation, differentiation, and maturation. After printing a 3-D structure using bioink, it is then stabilized by thermal, ionic, or chemical cross-linking methods. There are several hydrogels that are commonly used in bioinks and they can largely be classified into three kinds: natural-derived, synthetic, and semisynthetic materials [6]. Decellularized extracellular matrix-derived materials are a promising source of natural-derived bioinks that provide excellent biocompatibility and biofunctionality. Organ printing is being applied in the tissue engineering field with the purpose of developing tissue/organ constructs for the regeneration of both hard (bone, cartilage, osteochondral) and soft tissues (heart). There are other potential application areas including tissue/organ models, disease/cancer models, and models for physiology and pathology, where in vitro 3-D multicellular structures developed by organ printing are valuable [6].

3-D Printed Tumor Models: State of the Art

A great impact of 3-D printing upon tumor model generations is emphasized recently in this new approach [5, 6]. They are designed to mimic some of the key features of the in vivo tumor environment, giving rise to performance of the studies on the tumor initiation, progression, metastasis and even therapy and toxicity [6, 7]. The first 3-D tumor model was the one of *mammary gland* designed by Bissell in 1990s [8]. The best model so far is bone *Ewing's sarcoma* which is widely comprehended from different point of views [5–8]. Novak-Vunjakovic and Villasante consider microenvironment of the tumor in creating biologically relevant tumor models. It is a complex view, since tumor is not only CSC, but bulk of cells in specific environment that makes it tumor. In a nutshell, knowing tumor microenvironment and engineering a controllable biomimetic system that resembles a set of particular properties, using the bone cells and specific molecules of extracellular matrix on relevant scaffold, one can 3-D print bone tumor model to be used for examination of given properties [5–8]. Analogous to the "organs on a chip," engineered on microfluidic platform (to study physiology and evaluate drugs). Villasante and Vunjak-Novakovic proposed "minimally functional unit" that provides a limited but sufficient level of complexity for studying a specific aspect of the tumor [6]. Teams require integrated knowledge of several disciplines such as engineers and biologists.

A tissue model Ewing's sarcoma is produced through 3-D printing and incorporates bone derived from MSCs, ECM molecules, and tumor compartment with Ewing's sarcoma cells. That means, that cocultured tumor aggregates with MSCs were differentiated into osteogenic lineages within decellularized normal bone used as a scaffold [5]. Thus, tumor is residing within its native bone niche, while the natural cross-talk between cells and molecules, within the model, is going on. The entire process is followed with gene-related regulation (~12 genes involved in cancer, e.g., focal adhesion and cancer pathways, are reexpressed in this 3-D model) [5]. This finding is also significant from therapeutic point of view (gene-related; Fig. 11.1).

Drug toxicity is one of the most serious problems in anticancer therapy. It is not so seldom that patients in tumor cachexia die due to toxic effects of the anticancer therapy. The engineered tumor system can be very useful in both drug discovery and testing [8]. These models are just emerging [7–10] as a novel approach to cancer drug discovery and treatment options, demonstrating already their high potential.

For effective testing, it will be necessary in the future to incorporate capillary bed, stromal cells, the immune system components and physical signaling, all those aspects of tumor physiology that cancer stem cells have in their "cave of nativity"— niche. It just underscores the significance of Bissell's 3-D tumor model which has revolutionized the field of cancer research showing the necessity for further development of these concepts in order to understand and be able to control tumor behavior.

As we can see, the tumor modeling still uses cocultured tumor aggregates (not pure, isolated cancer stem cells) with normal MSCs. Using Cancer Stem Cells within these models we should be able to better support the basic concepts of the CSC theories and test their validity with high accuracy and precision.

Fig. 11.1 The 3-D printing

References

1. Jaswal S, Akashi K, ILW (2001) Nature 414:105–111
2. Bonnet D, Dick JE (1997) Human acute myeloid leukemia is organized as a hierarchy that originates from a primitive hematopoietic cell. Nat Med 3:730–737. doi:10.1038/nm0797-730
3. Li C, Heidt DG, Dalerba P et al (2007) Identification of pancreatic cancer stem cells. Cancer Res 67(3):1030
4. Li C, Lee CJ, Simeone DM (2009) Identification of human pancreatic cancer stem cells. In: Yu JS (ed) Cancer stem cells. Humana Press, New York, pp 161–173
5. Al-Hajj M, Wicha MS, Benito-Hernandez A, Morrison SJ, Clarke MF (2003) Prospective identification of tumorigenic breast cancer cells. Proc Natl Acad Sci U S A 100(7):3983–3988
6. Lee JSJ, Pati F, Jung JW, Jang J, Park JH, Cho D-W (2015) Organ printing 2015—to be published
7. Villasante A, Vunjak-Novakovic G (2015) Bioengineered tumors. Bioengineered 6(2):73–76. doi:10.1080/21655979.2015.1011039
8. Villasante A, Vunjak-Novakovic G (2015) Tissue-engineered models of human tumors for cancer research. Expert Opin Drug Discov 10(3):257–268
9. Kimlin LC, Casagrande G, Virador VM (2013) In vitro three-dimensional (3-D) models in cancer research: an update. Mol Carcinog 52:167–182
10. Villasante A, Marturano–Kruik A, Vunjak–Novakovic G (2014) Bioengineered human tumor within a bone niche. Biomaterials 35:5785–5794

Chapter 12
Summary on the Role of Bioengineering in Cancer Stem Cell Paradigm

The saddest aspect of life right now is that science gathers knowledge faster than society gathers wisdom.

Isaac Asimov

Stem Cell Paradigm: Resume

Stem cells are defined as cells that have the ability to perpetuate themselves through self-renewal and to generate mature cells of a particular tissue through differentiation. In most tissues, stem cells are rare. As a result, stem cells must be identified prospectively and purified carefully in order to study their properties. It is quite clear today that we have two essential groups of stem cells that we can potentially use for cellular therapy: ES and Adult stem cells. Both are unspecialized and both are effective in treatment of life-threatening diseases (Fig. 12.1).

Definitive experiments are needed to determine whether the cells from the bone marrow that are capable of giving rise to different non-hematopoietic lineages are indeed HSCs or another population [1]. If further studies support the idea of HSC plasticity, this will undoubtedly open new frontiers for understanding the developmental potential of HSCs, as well as expand their therapeutic application [1–5].

It is also obvious from this book that stem cell biology could provide new insights into cancer biology. There are definitely three aspects of the relationship between stem cells and cancer cells: first, the similarities in the mechanisms that regulate self-renewal of normal stem cells and cancer cells; second, the possibility that cancer cells might arise from normal stem cells; and third, the notion that cancers might contain "cancer stem cells," rare cells with indefinite proliferative potential that drive the formation and growth of tumors/cancers [6].

The stem cell paradigm for cancer adds a level of order to an anarchic process. Two routes from a transformed cell to a heterogeneous tumor (stochastic and clonogenic) were mentioned.

© The Author(s) 2015
M. Pavlovic, B. Balint, *Bioengineering and Cancer Stem Cell Concept*,
DOI 10.1007/978-3-319-25670-2_12

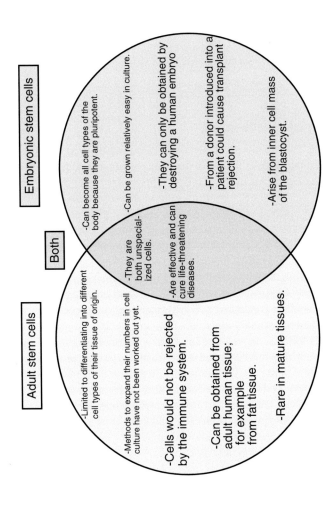

Fig. 12.1 Similarities and differences between

Self-Renewal of Stem Cells

One of the most important issues in stem cell biology is understanding the mechanisms that regulate self-renewal, crucial to stem cell function, because it is required by many types of stem cells to persist for the lifetime of the animal. Moreover, whereas stem cells from different organs may vary in their developmental potential, all stem cells must self-renew and regulate the relative balance between self-renewal and differentiation. Understanding the regulation of normal stem cell self-renewal is also fundamental to understanding the regulation of cancer cell proliferation, because cancer can be considered to be a disease of unregulated self-renewal. Perhaps the most important and useful property of stem cells is that of self-renewal. Through this property, striking parallels can be found between stem cells and cancer cells: tumors may often originate from the transformation of normal stem cells, similar signaling pathways may regulate self-renewal in stem cells and cancer cells, and cancer cells may include "cancer stem cells," rare cells with indefinite potential for self-renewal that drive cancerogenesis. Thus, stemness is still the issue that requires a lot of attention and research in order to give better ideas on how to approach to CSC (Fig. 12.2).

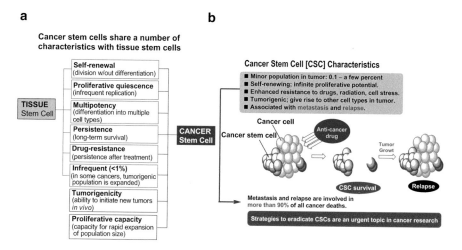

Fig. 12.2 (**a, b**) Properties of cancer stem cells

The Role of Bioengineering in Understanding CSCs Paradigm: Resume

It is the bioengineering that helped clarification of many issues linked to normal and cancerogenic stem cell paradigm. Development of equipment and techniques capable of detecting and isolating these cells was eruptive within last decade [1–20]. In the hope that this book has given a view on essential novel approaches and techniques needed to manipulate (isolate, purify, 3-D print, etc.) CSCs we are indicating only that the development of novel technologies, such as genetic engineering, nanotechnology, antimitochondrial cancer therapy, magneto therapy, immunotherapeutic approaches, and remote control drug delivery systems, will not only facilitate diagnostics and therapy, but also understanding of entire concept of CSC at molecular and other possible levels.

The Concept of Targeted CSC Therapy: Resume

It will not be that easy to target CSC although the concept seems to be very simple (Fig. 12.3). It might be necessary to figure out different CSC profiles and find the best possible targeted treatment for each particular CSC. Thus, the fact that only a few cells within the tumor, the CSCs are tumorigenic and possess metastatic phenotype has implications for both therapy and research. They must go in parallel in

Fig. 12.3 The concept of targeted CSC therapy visualized

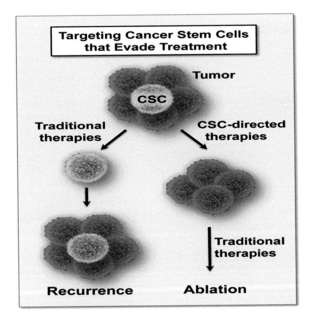

order to reach ultimate goal. Today, we can see different directions in approaches to Targeted CSC treatment, which is result of accumulation of knowledge and work over a hundred years and maybe even more [1–20].

Which direction to go? Is now, when we know CSC markers immunotherapy with autologous dendritic cells (DC) more attainable? Now, as we have fixed molecular marker as a hallmark of the cell that does not change as an antigen-will it be the solution? Could the vaccine against cancer be applicable to all of them and the best possible solution in each particular case? Being aware that Warburg's effect is prominent sign of malignancy, after so many years of persistent work of Racker and his followers, shall we go in metabolic trajectory direction? If these cells have less mitochondria than normal cells, and therefore are in hypoxic condition, in danger of cell death, shall we use the inhibitor of respiration or inducer of programmed cell death? Does magnetotherapy help in all cases of cancer stem cells or it is only in the brain tissue? Is the concept of nanorobot applicable in this case and how to create it for this purpose? Can we really engineer the normal stem cells to produce applicable toxins against cancer stem cells of all tissues?

One is sure: 3-D printing is advancing in designing tumors out of the body which will definitely help us to understand many aspects of the tumor development and growth and particularities, deeper and more complete. It is not the end of the research, but might be a very good beginning. Meanwhile, one can also try to improve bioinformatics tools in order to collect, analyze, and predict the possibilities in tumor and CSC biology that we can expect and react upon.

There are so many questions to be answered and there is a good time for that. Bioengineering is an explosively expanding field with the eyes open to the future.

References

1. Pavlovic M, Balint B (2013) Stem cells and tissue engineering. Springer, New York, ISBN: 978-1-4614-5505-9 (eBook)
2. Ratajczak M, Ratajczak J, Shin DM, Wan W, Liu R, Masternak MM, Piotrowska K et al (2011) Higher number of stem cells in the bone marrow of circulating low Igf-1 level LaronDwarf novel view on Igf-1, stem cells and aging. Leukemia 25:29–733
3. Balint B, Todorović M, Jevtić M, Ostojić G, Ristanović E, Vojvodić D et al (2009) The use of stem cells for marrow repopulation and in the field of regenerative medicine. Mak Med Pregl 63(Suppl 75):Str. 12
4. Mayfield J, Pavlovic M (2014) Current modalities and the implications of cancer stem cell engineering in oncological treatment. ART 14(1–2):49–59
5. Li C, Lee CJ, Simeone DM (2009) Identification of human pancreatic cancer stem cells. In: Yu JS (ed) Cancer stem cells. Humana Press, New York, pp 161–173
6. Reya T et al (2001) Stem cells, cancer, and cancer stem cells. Nature 414(6859):105–111
7. Pavlovic M, Balint B (2006) The use of stem cells to repair the cardiac tissue. Anest Reanim Transfuziol 34:129–150
8. Li C et al (2007) Identification of pancreatic cancer stem cells. Cancer Res 67(3):1030–1037
9. Neve RM et al (2006) A collection of breast cancer cell lines for the study of functionally distinct cancer subtypes. Cancer Cell 10(6):515–527

10. Al–Hajj M et al (2003) Prospective identification of tumorigenic breast cancer cells. Proc Natl Acad Sci U S A 100(7):3983–3988
11. Singh SK et al (2003) Identification of a cancer stem cell in human brain tumors. Cancer Res 63(18):5821–5828
12. Orlic D, Kajstura J, Chimenti S et al (2001) Bone marrow cells regenerate infarcted myocardium. Nature 410(6829):701–705
13. Pederson LP (2007) Warburg, me and Hexokinase 2: multiple discoveries of key molecular events underlying one of cancer's most common phenotypes, the Warburg Effect: i.e., elevated glycolysis in the presence of oxygen. J Bioenrg Biomembr 30:211–222
14. Fulda S, Galuzzi L, Kroemer G (2010) Targeting mitochondria for cancer therapy. Nat Rev Drug Discov 9:447–464
15. Huang X, El–Sayed IH, Qian W, El–Sayed MA (2006) Cancer cell imaging and photothermal therapy in the near–infrared region by using gold nanorods. J Am Chem Soc 128:2115–2120
16. Burke AR, Singh RN, Carroll DL, Wood JC, D'Agostino RB et al (2012) The resistance of breast cancer stem cells to conventional hyperthermia and their sensitivity to nanoparticle-mediated photothermal therapy. Biomaterials 33:2961–2970
17. Atkinson RL, Zhang M, Diagaradjane P, Peddibhotla S, Contreras A et al (2010) Thermal enhancement with optically activated gold nanoshells sensitizes breast cancer stem cells to radiation therapy. Sci Transl Med 2:55–79
18. Phillips TM, McBride WH, Pajonk F (2006) The response of CD24(−/low)/CD44+ breast cancer-initiating cells to radiation. J Natl Cancer Inst 98:1777–1785
19. Galanzha EI, Kim JW, Zharov VP (2009) Nanotechnology-based molecular photoacoustic and photothermal flow cytometry platform for in vivo detection and killing of circulating cancer stem cells. J Biophotonics 2:725–735
20. Galanzha EI, Shashkov EV, Kelly T, Kim JW, Yang L et al (2009) In vivo magnetic enrichment and multiplex photoacoustic detection of circulating tumour cells. Nat Nanotechnol 4:855–860

Printed in the United States
By Bookmasters